A Companion of Feminisms for Digital Design and Spherology

Amanda Windle

A Companion of Feminisms for Digital Design and Spherology

palgrave
macmillan

Amanda Windle
London College of Communication
University of the Arts London
London, UK

ISBN 978-3-030-02286-0   ISBN 978-3-030-02287-7   (eBook)
https://doi.org/10.1007/978-3-030-02287-7

Library of Congress Control Number: 2018957450

Cover illustration: © Melisa Hasan

This Palgrave Pivot imprint is published by the registered company Springer Nature Switzerland AG
The registered company address is: Gewerbestrasse 11, 6330 Cham, Switzerland

Acknowledgements

Alison Marlin encouraged me back in 2012 to write this book. Thanks to Alison along with Christine Battersby, Annika Joy, Lara Houston, Kris Cohen and Simon Willmoth for critically reading draft chapters; and for the book proposal, thanks Kat Jungnickel. Special thanks to Alex Taylor and Noortje Marres for helping review the book in addition to the anonymous reviewers.

I gained thoughtful feedback from Tom Gieryn and Lisa Messeri during 4S, in San Diego (2013), where I first wrote about GIS-mapping. And to Steve Cross for loaning me copies of Slot-machinen's texts. Thank you.

Thanks to the graduate students at the University of the Arts London (UAL) who have participated in discussions about Spheres, particularly Carl Grinter. Thanks to Helga Steppan's invitation to workshop a chapter with the visual communication third-year undergraduates at Linnaeus University, Sweden, especially exchange student Antara Madavane, from Srishti University. Thanks Ivor Williams for your encouragement about my story on Venn diagrams and for welcoming me to Università Iuav di Venezia (IUAV) where I began drafting that chapter. Thanks to Nina Wakeford for supporting me through other short-form writing—and whose advice carried into finishing this book. Thanks to Janet McDonnell for mentorship. Thanks also to Maurn Turner for psychotherapeutic listening.

The research for this book has been funded by Microsoft Research Ltd., the Higher Education Innovation Fund (earlier in my role as the DigiLab Fellow), and the Higher Education Funding Council for

England, for the roundtable event. Thank you to Lucas Joppa, Kristin Tolle, Kenji Takeda and Matthew Smith for all your generosity on the Threat Mapping design interface project at Microsoft Research Ltd.

The departments of research and international at UAL have given me countless travel grants to get parts of this book peer-reviewed at 4S and EASST in Barcelona, Boston, Buenos Aires and San Diego. Thanks also to UAL's ethics committee, legal team and library, and to the Research Management and Administration team because their work often goes un-thanked.

Many thanks to the editors and co-editors at Palgrave Macmillan: I'm indebted to Holly Tyler, Joanna O'Neill and Sophie Li, but particularly to Josh Pitt for writing out thoughts about peer-review for Backchannels (4S). Thanks to Jenn Tomomitsu for copy-editing pre-manuscript submission.

Thanks Annika Joy and Andrew Hill for your friendship. Thanks Ranjita Dhital, Geoff Morrow, Ian Storey, Silvia Grimaldi and Nela Milic for encouragement. Thank you Jason Rainbird for your love, design companionship and for accompanying me whenever I need to go hike a volcano.

CONTENTS

ABBREVIATIONS

LIST OF FIGURES

LIST OF TABLES

CHAPTER 1

Finding a Spherology of Feminisms for Designing Media

Abstract In this chapter, I introduce a strategy for working both *with* and *against* Peter Sloterdijk's *Spheres* trilogy ([1998–2004] 2011–2016). The spatial questions of *Spheres* are related to being-in the depressive sphere, which is better known temporally as the Anthropogenic epoch. This book brings new practice-based contributions to reading *Spheres* by providing design-ins (or entry-points) which act as wayfaring for design readers. By reconfiguring the feminist writers that are referenced in *Spheres*, I put forward a renewed strategy—that is, a spherology of feminisms. With new approaches to *Spheres* through embodied situations, methods and technological problems, this companion is relevant to the spatio-temporal aspects of design in general like studio practice, design layout or digital interaction.

Keywords Being-in · Design-in · Feminisms · Spherology · Technoscience

While reading and re-reading a three-part trilogy called *Spheres* by Peter Sloterdijk ([1998–2004] 2011–2016), I got on with designing. I took eight years on and off, taking longer to read Spheres than it probably took the author to write it. I stayed for a while because it's a complex trilogy and I continued to read feminist-queer-crip-science fiction too. I journeyed into the literature recommended by *Spheres* and read that too. I read myself into writing a companion. But I did not plan to.

© The Author(s) 2019
A. Windle, *A Companion of Feminisms for Digital Design and Spherology*,
https://doi.org/10.1007/978-3-030-02287-7_1

There's so much ground covered in the trilogy that many readers will experience an inspiring literary read that helps creative practitioners to laterally think, make and do. Unintended by Sloterdijk, I experienced a renewed interest in feminist technoscience. *Spheres* was published a decade after Donna Haraway first wrote the "Cyborg Manifesto" in 1983, and Octavia Butler had written the science fiction trilogy, *Lilith's Brood* between 1987 and 1989. I read these books before I read *Spheres*. I didn't discover feminist technoscience through *Spheres*, but it sure seems like Sloterdijk was doing just that in the final volume of *Spheres*. The broader word feminism concerns me in *Spheres* because it's not the generative and caring set of literatures that I know. The main offering of *Spheres* is to suggest that we live in co-isolated clubs and the trilogy does this by bringing together a range of disciplines. By erratically gathering contexts to create a continental philosophy, Sloterdijk has inadvertently gathered a plurality of feminisms, but it's no club. The feminist ideas are dislocated and sidelined, but they're also intriguingly multiple and divergent.

As a designer rather than a philosopher, I share Sloterdijk's implicit fear of global failure as an impetus to write about the depressive sphere. Sloterdijk is not a designer but a philosopher and Chair of Media, at ZKM, Zentrum für Kunst und Medientechnologie in Karlsruhe, Germany. While the trilogy refers spatially to the "depressive sphere", a reader may be more familiar with the temporal term of the Anthropocene, an epoch named at the International Geological Congress in 2016, where a feminist technoscience reader will know Donna Haraway's term the "chthulucene", which emphasises the biological taxonomy of multispecies (2016). To produce and consume during this geological time is in itself an act of designing our shared atmosphere.

The trick to reading *Spheres* is to bring more to it so as to implicate design into what being-in the depressive sphere might mean (I, 468). Digital "atmosphere designers" (III, 65), of which I am one, work on problems like designing technical systems in the "depressive sphere" (I, 468). Sloterdijk figures atmosphere designers as architects, graphic designers and engineers but more broadly to how all humans can be considered as such through their product consumption. I contribute to this term by including sentient machines as atmosphere designers too (in Chapters 3 and 6). I also entangle the three prevalent metaphors of *Spheres* which are bubbles, globes and foams. Bubbles signify intimate human relations between beings and things. Globes relate to maps and global systems like economic markets. Foam(s) are the co-isolated social arrangements like being-in an apartment block. Foam(s) is the trope most practical for considering what is defined socially as atmosphere design.

While I find *atmosphere designer* and *depressive sphere* generative terms, it became clear to me that my way of thinking about this agential situation had been fragmented and interrupted by Sloterdijk's narrative voice. Reading and re-reading about dislocated female bodies in pain and stress requires a reparative approach. How women's bodies are used to explain the abstract and intangible aspects of the social environmental sphere and how they continue to pervade in design philosophy and practice needs further enquiry. Without such, the reader may continue to spiral down into *Spheres* without ever really knowing why, while those in the know, i.e. whose disability, impairments and elliptical viewpoints support their understanding of gender, may have strategies to deal with texts that are written for an abled-bodied reader. I take courage from authors who write about bodies from experience. I gain strength from writers like Sarah Lochlann Jain (2013), Jackie Orr (2006) and Celia Roberts (2007). I will not explicitly work with these texts, but I will highlight them upfront for differently abled readers.[1] And, my body will crop up in the text because bodies are one particular boundary for registering the depressive sphere.

Feminists do work at the centre of ideas, but I intend to bring the peripheral ideas of *Spheres* to the surface because that's where the spherology of feminisms is situated. Scholars choosing to cite from *Spheres* need to understand how Sloterdijk makes his argument. To draw on the feminist writer, bell hooks, "a main body [is] made up of both margin and center" (2000, xvi). While a body refers to a physical body, I relate this idea to the body copy of a book and to embodied reading practices.

[Here I am, cropping-up again] I grew up talking like a cyborg in a human world. English, Sicilian and Portuguese were spoken at home with accents that I hear well, but the meanings of these languages I understood predominantly through affect, body language and intonation. I learned German, but my ability to speak it is also limited. I explain this so that the reader understands that I waited and read *Spheres* in English and then checked it in German. So, when I read the final volume in German (not the parts pre-published in CCR), I read the index first. This deficiency became pivotal in writing this companion. If there was a centre to my account of spherology, it would be in the indexicality of *Spheres* (i.e. the index, citations, illustrations and endnotes) discussed mainly in Chapter 4 (a centre chapter).

New translations have sustained popularity in *Spheres* mainly from an academic audience consisting of art, design, media, theology, political and social sciences leading to doctorates through to journal papers, as well as edited books and monographs. The trilogy was initially published volume-by-volume in German (1998, 1999 and 2004) before being translated into English (2011, 2014 and 2016). There are texts that canonise Sloterdijk's work like Jean-Pierre Couture's philosophical reader (2016) and Stuart Elden's edited collection for scholars in the humanities and social sciences (2012). There is, however, no companion that focuses on *Spheres* for a design reader, or for its history of media ideas and rethinks the concept of companion-writing which I aim to address. Companion texts like readers and edited collections are usually the final part of canonising literature, but *Spheres* got a second market thanks to Bruno Latour, Sloterdijk's friend and scholar of STS (studies of science, technology and society). For earlier readers of *Spheres*, they can unlearn how Sloterdijk became world-leading and ponder citation impact.

I am neither a friend nor colleague to Peter Sloterdijk. But I write as a stranger, familiar to Spheres. I say this to stress that a designer need not take up spherology by his name alone—to join a 'Sloterdijkian' club—in the ways that academics promote a name. To be concerned with a co-isolated club of feminisms embedded in *Spheres* is to rethink what is "patterned after a male model" (hooks 2000, 149) and to show how spherological feminisms uncomfortably orbit *Spheres* as obscured, forgotten, de-centred and marginalised texts. Feminisms are the way to get around the manipulation of *Spheres*.

A reader may choose never to read Sloterdijk, but this would be an extreme choice I don't support. However, it may be necessary to skip *Spheres* altogether for those who feel their reading deeply. Reading *Spheres* affected my health, but it was also healing research. I wrote this book with an autoimmune disease which is an autobiographical detail I've added so that the reader can understand that I write as a woman with an invisible condition of interiority—of impairment-ill-health-disease-disability. I write in a different register to Sloterdijk because he neglects yet creates a text for a differently abled reader. My companion is not a 'crutch' to *Spheres* but a way of writing-out with a body that reflects an epoch of systemic dis-ease. I counter Sloterdijk's writing *about* the immune system by writing explicitly *with* and *through* my chronic

life experiences because the genetic aspects of *Spheres* range from trying
to offensive. When necessary, I've added excerpts of autobiography like
a discussion of anaemia in Chapter 5 and references to collective care
in Chapter 6. To an extent, my companion is about surviving *Spheres*,
because to take it seriously can mean taking offense, and that's stressful.
It's conflicted work that I felt necessarily able to do. And, my account of
design is advocating for more design researchers to claim their disability and impairments in their working lives and if inclined to write about
designs' shared inwardness.

BEING-IN EXILE: REMOVING *GESTELL* (FRAME) TO EXPLORE *DASEIN* (BEING-IN)

Many of Sloterdijk's stories in *Spheres* are unreservedly about disembodied
female bodies as prosthetics that prop up a masculine whole, something of
which I'd argue is critiqued in Fig. 1.1. The design STS academic, Nerea
Calvillo, has provided a collaborative methodology towards a designed
condition of interiority. Having a riot of music fans secured in walls of a
polyvaginal space is an interesting S.O.S. and a particular design critique
of inwardness. Simultaneously, Calvillo is working both *with* and *against*
a condition of interiority shared through music practice. There's a balance
to be made between working with design interventionist strategies that
can simultaneously intervene with a problem because doing so can draw
attention to the problem it chooses to diminish.

Feminist spherology and *Spheres* philosophy share one aim: to figure
statements about Anthropogenic interiority through the body. However,
from that shared aim, the spherology of feminisms found in *Spheres*
diverges. The reader can use this companion recursively to go back and
unpack the ethos of *Spheres* which is this: Sloterdijk didn't invent a feminist spherology or a spherology of feminisms; what he did was dismantle
and exile it to the endnotes.

What's centre stage in Sloterdijk's spherology isn't philosophical per say,
but an enframing of anthropological, artistic, designerly, environmental,
literary, sociological, theo-critical and spiritual contexts. He brings media
theory to these disciplines and relates them to a phenomenological concept
of 'being', particularly the human-centred position of 'being-in' which
is derived from Martin Heidegger's philosophical notion of Dasein—
meaning *a being-in-something* (I 89, 271 and 333–42).

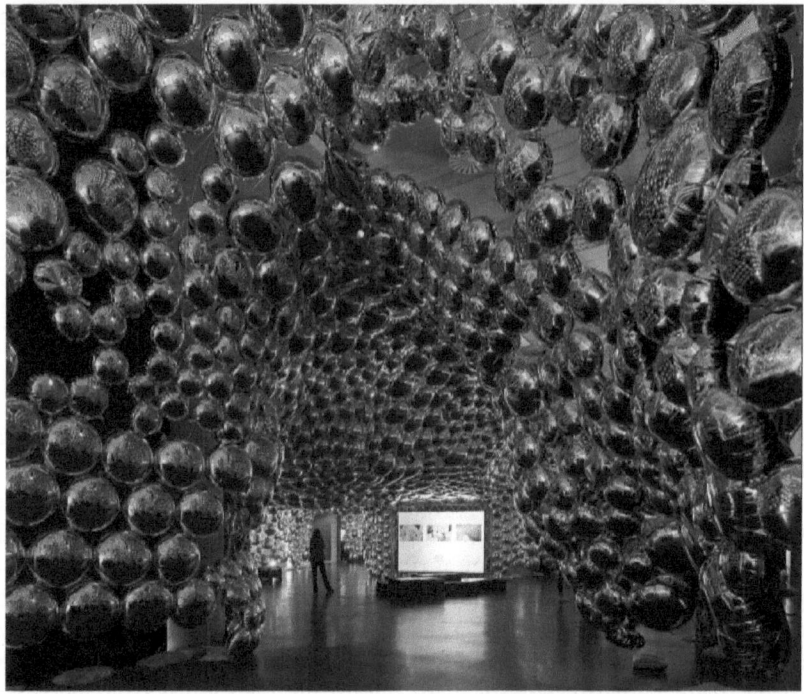

Fig. 1.1 Exhibition of the Polyvagina de Fan Riots; a flexible skin that creates "una condición de interioridad"; a "condition of interiority" to host Fan Riots; an artistic event curated by Ivan Lopez Munuera for the SOS music festival. Design: C+arquitectos: Nerea Calvillo with Marina Fernández. Assembly: Workshop with students of architecture of the University of Alicante led by Miguel Mesa del Castillo. Photographers: Miguel de Guzmán and Rocío Romero (*Image Source* http://imagensubliminal.com/polivagina-de-fan-riots/?lang=es. Last Accessed 3 Aug 2017)

> Being means someone (1) being together with someone else (2) and with something else (3) in something (4). This formula describes the minimum complexity you need to construct in order to arrive at an appropriate concept of world. Architects are involved in this consideration, since for them being-in-the-world means dwelling in a building.
>
> —Sloterdijk TTM, 7

One way to interpret *Spheres* is that it's Sloterdijk's brave attempt to write an autobiography without resorting to disclosure. It's either

Sloterdijk's philosophical recovery from being-in the Rajneeshee cult, or an extension of it *through* philosophy. If approached in the former way, a reader may not get drawn into exploring the latter explication (logic of production), which I'll now briefly contextualise.

Sloterdijk places spherology where it has always been—at the centre of stories that lead to affirming patriarchy. Being-in *Spheres* is uncomfortable when reading about the violence towards women as a masculine perspective on female sexual liberation because it's never a very balanced perspective. For every mention in *Spheres*, all the way to Osho—the Indian mystic previously known as Bhagwan Shree Rajneesh—the actions of his closest ally and assistant secretary, Ma Anand Sheela, are absent. Sheela's part in a planned bioterror attack and their prison sentences for wiretapping and fraud and subsequent deportations are remiss. I wonder about Sloterdijk's role in the Rajneeshee cult, particularly when encountering a photograph (from Sloterdijk's personal archive) of group therapy in Pune, India (II, 201).[2] Did Sloterdijk or Sheela stand in the doorway to photograph the nude female bodies lying on the floor beneath the roaming group leader?[3] Rajneeshee therapy involved finding a catharsis through sexual violence, an idea taken up by Sloterdijk (see Oliveira et al. 1996). This means that the "cynicism", as Sloterdijk calls it (think cynicism for now, read CCR or go straight to Chapter 2), is as much derived from his experiences in twentieth-century India than it is from post-war German philosophy.

Imprisonment and exile are the missing echo-chambers of Sloterdijk's autobiographical voice. Here's one about media: for every mention of the pop artist, Andy Warhol, I think of the feminist activist Valerie Solanas (they were lovers), who shot Warhol. When Sloterdijk repeats that Warhol considers his tape recorder as his wife (see Chapter 3), I think of this desperate act—of Solanas disappearing further into self-harming exile (I, 401). Of all these figures of exile, it's Solanas and Sheela that are exiled from *Spheres* (with only Osho and Warhol named) which strengthens a positive and masculine account of sexual liberation.

I want to forewarn the reader about the seduction of Sloterdijk's humour because it's also entangled in Osho's "dirty jokes" which are permissible for those beyond enlightenment (Rajneesh 2018). Religious studies scholars, Irving Hexham and anthropologist Karla Poewe, suggest that beneath the humour of the Rajneeshees are "real psychological disturbances" (1986, 147) and that "many of them made valiant efforts through much of their lives to correct or master their personality disorders—an indication of courage, not narcissism" (ibid., 120).

Humour is Sloterdijk's way of coping with exile, and while cathartic, there's a lot of sidestepping. Incredibly clever prose and dense theory point the reader in any direction except towards the autobiographical centre of Sloterdijk's pain. If that's not enough, the humour finally lifts the reader up from the depressive depths of Sloterdijk's despair to deliver a punch that knocks out of the reader all questions about the author. It's smart and witty, but it's strikingly violent. Understanding Sloterdijk's humour is to ask "what is this humour removing"? The answer being— the pain of exile.

Many of the literatures canonising *Spheres* renders Sloterdijk as an ace of spaces. To me, he's not the ace of spades and I do not write as an *a*-muse-(d)iva-cunt-Queen-of-Sheba. As a provocateur, Sloterdijk appears at times in his writing as a fascist tourist masquerading as an anthropological-priest in exile. Sloterdijk does have something important to say to men (who were born male and remain male) about exile, but *Spheres* won't help any person of gender (fluid or binary) to get there. He hints at a crisis in fatherhood in just over a paragraph of the first volume (Chapter 5) which tags onto a lengthier discussion of motherhood. Karla Poewe's auto-anthropology reveals that growing up in Germany during and after the Second World War meant being-in a predominantly female domestic and public sphere (2018, 7–8). Sloterdijk could be writing-out a similar life-defining experience.

In the third volume, Sloterdijk denies the personal sphere altogether to make sociological points about shared loneliness. He does this in first-person in the tradition of autobiography and literature, yet in the closing pages of *Spheres* his authorial voice splits into intra-disciplinary voices of the macro-historian, literary-critic and theologian (III, 801–827). It's how Sloterdijk's narrative about a society that is individualised, primal and exploited ends. It predicts human exile from planet earth, but does so as a philosophy that subsumes history, literature and theology. It's a text that can overwhelm a readers' disciplinary attachments, especially to design and media. My companion fills in some of the gaps in *Spheres* by refusing Sloterdijk's justification of Heidegger's enframing which is tucked away at the end of the trilogy (III, 835). While I'd agree with enframing in that "facts" are often artificial, constructed and forceful (which creates new absences in the process of doing research), I disagree with Sloterdijk's use of enframing because it enables him to write over-zealously.

That's just about as far as I'll go in this companion with expanding a Rajneesheeian perspective of *Spheres*. First, it's intrusive thinking and more along the lines of a journalistic biography rather than media and/or design research. Second, there's a limit to me trying to empathise with Sloterdijk's pain when it's undisclosed. And, third it's more than Sloterdijk will ever give to understanding the actions of women like Sheela and Solanas. Instead, I unpack how *Spheres* relies on colonial attitudes towards the narration of ill-health because Sloterdijk was a part of an international and interethnic cult whose members were often wealthy and part of a mobile, upper-middle class (Hexham and Poewe 1986, 120). There's a need to confront the standpoint that privileges male voyeurism in stories of female despair but not by proposing the opposite because for Rajneeshees (like Sloterdijk), sex and violence involve becoming-male if female, and vice versa. Meaning, for Sloterdijk, until the "whole universe has become woman".[4] Yet, *Spheres* isn't an act of meditative bi-empathy, its biuneness (coupling) is something else entirely.

The task that lies ahead is to test the extension of *Spheres'* theory for a wider range of designers working digitally by providing a resource of feminist ideas and methods not prevalently taught in design schools today. I do so because there's often a shortage of know-how rather than willingness. I go further than offer a slight twist on Sloterdijk's theory by cynical means because it's been "done to death" (pun intended). Future spherologists may want to take up a differently abled spherology. Approaching *Spheres* in this way shows where Sloterdijk's ideas lose their leverage and hints at what design researchers may need to be doing today: questioning the occult in design.

> When someone tries to 'agitate' me in an enlightened direction, my reaction is a cynical one: The person should get his or her own shit together. [...] For I do not like being asked, 'Then why don't you do something?'
> —Couture repeating Sloterdijk in CCR, 89

A reader doesn't always need to get their shit together. When reading about an author like Sloterdijk who is, in fact, well-versed and holds a doctorate in philosophical autobiography, it does lead the reader to diffractively ask both biographical and autobiographical questions of themselves.[5] That's why I don't entirely agree with the quote above. Reading *Spheres*, let alone attempting to "write-out" *Spheres*, will make any reader question their identity (Cixous 1976), which is why I'd argue that a companion of feminisms is necessary.[6] Both designers and philosophers may push away the autobiography found in Sloterdijk's work, but this is a mistake. There are a few places for autobiographical thinking in design and one of those opportunities is reading texts like *Spheres*. If designers don't contemplate who they are then how can they possibly understand how to design in the depressive sphere? Autobiography poised at designers needs a plurality of rotational approaches like reflection (Schön 1983), self-recognition (Haraway 2016) and diffraction (Barad 2003, 803), which will be introduced as required in this companion.

THE DESIGN-INS

My book consists of entry-points or rather "design-ins" to help designers reflect on technological explications. They're provided because *Spheres* doesn't relate directly to the embodied experiences and material practices of design. This section is an overview of what I mean by design and the chapters ahead, but first, I start with a note on writing.

My writing can be disorientating and it's sometimes difficult to find the core of a chapter because I don't always write that way. I draw on the rhetorical styles of *Spheres* and write recursively so this companion won't be any less difficult to read than *Spheres*. I used to think it's important to write clearly and confidently in order to produce *well*-rounded thoughts, but this runs counter to my experience. Perhaps we are all interrupted, but some of us have been interrupted more than others. My life has been full of interruptions and my writing and thoughts have taken shape through challenges with autoimmunity, militarist and masculine violence and resistance to becoming a muse. It seems apt to write interruptedly about the borderlines of depressive situations.

I moved into design for several reasons. There were few women occupying fine art academic positions in the 1990s and I later found this an almost impossible route of employment in higher education as a first-generation academic. Second, I was frustrated with the alienation of the audience in the conceptual art exhibited in white cubes and collected by the advertisers, Saatchi and Saatchi. I moved into design because it promised to be both participative and collaborative. However, what I found prevalent in British university education this side of the millennia was something more individualistic—that is, critical-speculative design.

Speculative design as derived from the British designers, Tony Dunne and Fiona Raby, critiques digital and electronic technology and their industries (2001). However, the influence of fine art critique on design methods in their work often gets ignored, no doubt because of the hierarchies that have been at play for centuries. Are these hierarchies now in reverse? Note the use of the shared pseudonym for the Beggarstaff brothers, William Nicholson and James Pryde in the 1880–90s. They used a pseudonym to protect their reputation as fine artists from earning a living through design (Bronkhurst 1978). I'm not arguing that today's designers reinforce this hierarchy of practice, but I want to point out that there are many reasons why the relationship between fine art and design became fused. Dunne and Raby trained in industrial design but the former also studied fine art. Their co-publications do not emphasise fine art in their practice, but it's implicit. Referencing to fine art is only evident in Dunne's doctorate which draws fine art enquiry *into* industrial design matters (1997).[7] Raby has asserted that their shared practice is about using humour to critique industrial design emphasising British wit (2012).

A more conventional design companion would lean on Sloterdijk's Indo-German humour and strengthen links between speculative-critical design and *Spheres*. I could have drawn on architecture, graphic and industrial design critiques and added a focus on digital design through fine art. But it's not enough in the late twenty-tens to critique from these sidelines. I'd argue that design and technology in academia are not always challenged enough through other modes of knowing. Critical-speculative design is one explication, but I garner other insights from leading design projects, gathering methods from design ethnography, graphic design, interaction design, human-computer interaction, performance design (scenography) and data visualisation techniques.[8]

I take up a broader set of design methods to circumnavigate strong claims made *for* design. Take the next quote from the design-researchers Ilpo Koskinen's et al. (2011):

In Bruno Latour's philosophical language design things turn weak hunches into stronger claims. They also translate many types of interests into joined strongholds and provide tools that take design from short to long networks.

—Koskinen et al. (2011)[9]

What's missing in this account are the ways that Madeleine Akrich (with Latour and others) were trying to bring actor-network theory (ANT) together with "flow" methods (Akrich et al. 2002), and how they did so by studying innovation in research and development (R&D). Or how Bruno Latour, along with Pablo Jensen and others, attempted to bring together *Spheres* and ANT through a digital test involving digital interface technologies (2012). A reader doesn't need to know about Latour's ANT to read my book or *Spheres* for that matter, but these are well-written texts about network and flow relevant to digital R&D. Each time I introduce the contexts of new technologies, whether it's working remotely (Chapter 1), a backdrop of sentiment analysis (Chapter 2), ethical concerns of machine learning (Chapter 3), digital search capabilities in the indexicality of books (Chapter 4), data tracking and visualisations of the quantified-selfers (Chapter 4), a critique of international design standards (Chapter 5) or working with large vector-based databases (Chapter 6), it's feminist technoscience and embodied STS that I draw on to think through *Spheres*. Koskinen and the other all-male authors suggest that "Design research practice builds heavily on bodily and social interaction, which is difficult to do in the virtual domain" (op. cit., 125). Yet those with disability and impairments may say quite the opposite. It's difficult when literatures of design begin to split along lines of gender because it means that designer gets caught up in drawing polarised boundaries around design as a way to cope with the depressive sphere, which is what Sloterdijk is both doing and undoing in *Spheres*.

Now to introduce the chapters of this book, Chapter 2 explores an alternative to the premise that *Spheres* should be read as a novel. I draw on the ways a creative practitioner may read texts so as to introduce the cynicism of *Spheres* and its background in female protest. The chapter interweaves three generations of feminist writing. In recent years, practice-based research has focused on synergies between writing and making and in some areas of design STS, it takes place in studio spaces that perform like the ancient philosophical agora and less like spaces for queer-thinking and sexual-arrest (meaning spaces to take a break from white-cis-hetero-patriarchy). To explain this chapter another way, it's about how to read *Spheres* when doing digital research. As a design-led approach, I start with the marketing of *Spheres*.

Chapter 3 considers a reading group that takes place outside of studios, libraries and labs at the Royal Society of the Arts. The event gathered the very interdisciplinary audience this book intends to reach, drawing together a roundtable of art, design and STS academics. Instead of recounting the group debate, I expand on Sloterdijk's interest in foamed clubs. However, rather than focus on the mega-conference as Sloterdijk recommends, my take is on the intimacy of reading together. I explore how machine learning and digital labs enter into the ethics of atmosphere as an explication of data processing an audio transcription of the group debate.

Chapter 4 reviews the citation practices of the Sphere's index to look at the forgotten women in *Spheres*. The chapter questions the trilogy's structure and agency. While Sloterdijk's trilogy might be experimental in prose (philosophy as a novel), the index innovatively lags behind (excused by enframing). By gesturing towards some technical predictions around diversity I set up the provocation that book indexes matter. The trilogy is considered world-leading, but the index reveals something less diverse regarding disability, ethnicity, race, sexuality and gender. I rethink the indexicality of *Spheres* by offering a series of data visualisations and consider *being-in-a-diagram*, notably the Voronoi and Venn diagrams.

Chapter 5 focuses on the way images perform in the *Spheres* trilogy and through Sloterdijk's only design collaboration with Gesa Müeller von der Haegen and others in "Making Things Public" (2005). I begin with a mystical exploration of Gian Lorenzo Bernini's sculpture, Santa Teresa d'Ávila, which appears in the trilogy's first volume. Following this is a speculative scenography that investigates what design is doing within international politics, while working with wireframes. It's a baroque critique concerning sculpture, typography and page layout entangling perspectives from Britain, Germany, Italy, Mexico and Syria. It's also a way to look at the design standards in publishing and how visual ideas get credited in philosophy. The aim is for designers to understand know-how in terms of ideas and methods, through these parables of practice.

Chapter 6 investigates the logics of environmental map ping and the design and development of a geospatial information system (GIS), i.e. a web interface. The design project involves some of the simplest taxonomic arrangements of environmental threats to species and is a form of environmental science work evidenced through drawing. Hand-drawn and digital maps coexist for various environmental groups mapping albatrosses to antelopes, microbes to molluscs, lions to lizards and so on. In spherical terms, it's a *foam* issue. I review several explications to enquire "what is it like to be a digital 'atmosphere designer'"? I focus on a design error interweaving my embodied account of designing and testing a web-based application.

Generally, my book advocates for caring about the depressive sphere by looking after the embodied designer reading *Spheres*. What I bring out of *Spheres* is essentially a spherology of feminisms. It's a way to remove the debased ideology of Rajneesheeism, which are the clues to experiencing *Spheres* as a manipulative text. My alchemy was to find a spherology of feminisms so as to make design-ins for those who don't feel like they belong. I unpack the criticisms of *Spheres*, the design collaborations and the sexualised performances of its art and mysticism. This research is necessary so that readers know that it takes a bit more than just project management techniques, graphic design software and programming to design-in the anthropogenic-chthulucenic-depressive-sphere. Designers know all too well that climate change exists. The challenges for the design academy (whatever disciplines may figure into its future) are not just about recycling our materials and consuming less; they're about unpacking the occult in today's design methods as a mode of philosophical repair. I enjoyed forensically finding a spherology of feminisms in *Spheres* for the manipulation of the trilogy requires autobiography. A designer may use this companion to reflect on their own modes of design knowing to consider what anchors practice to the depressive sphere, and what gets exiled and why. I'll use parody as a way of incorporating humour. *Spheres* should be primed at designers because, as I'll repeat, Sloterdijk recognises design as the occult.

I got lost in the occult of *Spheres* (that's the point), but I found my way out of this design-in by reconnecting the co-isolated club of feminisms and finding attachments to disagreeable feminisms. It is *with* a multiplicity of feminisms that I promote care over violent catharsis. And by writing-out about togetherness, I'll argue for designers not to write-off autobiography as a mode of knowing. This, I assure, will help any discerning design reader develop their creative practice *with* and *against* *Spheres*. By bringing my physical impairments to bear witness to designing in the depressive sphere, I do so through knowing that healthy bodies cannot fully elicit what's needed in the anthropogenic epoch.

Do I have other *Spheres* issues? Sure, I take issue with the patriarchy of design and with gender violence enacted on the designed page. I have read *Spheres* in detail to understand being-in-depressive-digital-technoscientific-futures of now, but with the aim of illuminating to others the spherology of feminisms that exists before, within and beyond *Spheres*. Through writing-out this companion, I realise that designers need autobiography and shared queer-crip-feminisms to live well and design together. To an extent, this perspective means *staying-in* the troubling worldviews that *Spheres* curates. I cannot entirely advocate for distancing from particular worldviews completely as Haraway does in recent years (2016). Had I followed Haraway's advice which was until 2016 to sideline religion, psychoanalysis and disciplines that have done so much damage to women, then I wouldn't have really known *Spheres,* or the deathly risks of female emancipation framed through a patriarchy of sexualised liberation. A scholar cannot through the string figures of Haraway or labyrinthian stories of Sloterdijk alone know this mode of knowing, but they have very good reasons for not doing so, because it's to know the depressive as a collective, sped-up death march to**war**ds exile. By taking a detailed look back to Haraway's earlier work (among others), a compromise of working *with* and *against Spheres* is overdue. *Spheres* is a recursive design philosophy put painfully into my reflections of design practice.

NOTES

1. The phrase "differently abled" when referring to disability in North America is contested, considered euphemistic, and few persons of disability refer to it (Linton 1998). Thanks to neuro-ethicist Cynthia Forlini for pointing this out. However, I sometimes use it to help me navigate moments when disease and disability are in flux. I like the active fluidity of the term. Symptoms and diagnoses can be both simultaneously fluid and static when understood by the medical profession as parts of the body—of blood, cell, colon, heart, joint, kidney, lung, marrow, sac, skin, and womb; of endo-cardio-gastro-gynae-rheuma-specialisms etc. My autoimmune disease(s) have been diagnosed, misdiagnosed, and re-diagnosed. Being-ill and becoming-ill are different temporalities belonging to my body and they can be disabling. Being differently abled however, means that I do things differently, meaning—I pace my days differently to enable myself to be able.
2. Sloterdijk was photographed wearing Rajneeshee clothing in Amsterdam, 1983 (www.oshonews.com/2015/02/08/sloterdijk-cynicism/).
3. Wolfgang Dobrowolny secretly filmed these sessions (often leading to assault and rape) taking place in padded windowless rooms in Pune, India (1981). In *Globes (II)* the session is in a room with windows, and potentially by a different photographer.
4. See Rajneesh, "The Tantric of Erotic Love", Resurgence 5, May–June 1974, 11–16, in Hexham and Poewe (1986, 113).

5. *Spheres* is inspired by Robert Musil's, *Man without Qualities,* a novel completed by Martha Musil ([1943] 1966). Readers of my companion may find further inspiration in Gertrude Stein's autobiography, written in the guise of her life partner ([1933] 2001).
6. Sarah Kember's book, "iMedia" (2015) is an exemplary example of Hélène Cixous's method of "writing-out" and crucial reading for designers and readers of media (1976).
7. When I think of speculative design, I think of the Surrealism of Méret Elisabeth Oppenheim, for her cup, saucer and spoon made of fur, entitled "Le Déjeuner en Fourrure" (1936).
8. For a contemporary account of critical design see Matt Malpass (2017) and Daniela Rosner (2018).
9. "Design Research Through Practice" by Ilpo Koskinen's et al. (2011), refers to Latour's book entitled "Science in Action" (1987).

References

Akrich, Madeleine, Michel Callon, and Bruno Latour. "The Key to Success in Innovation Part I and II: The Art of Interessement." *International Journal of Innovation Management 6,* no. 2 (2002): 187–225.

Barad, Karen. "Posthumanist Performativity: Toward an Understanding of How Matter Comes to Matter." *Signs: Journal of Women in Culture and Society* 28, no. 3 (2003): 801–31.

Bronkhurst, Judith. "1895: The Beggarstaffs' 'Annus Mirabilis'." *The Journal of the Decorative Arts Society* 1890–1940, no. 2 (1978): 3–13.

Cixous, Hélène. "The Laugh of the Medusa." *Signs: Journal of Women in Culture and Society* 1, no. 4 (1976): 875–93.

Couture, Jean-Pierre. *Sloterdijk.* Cambridge: Polity Press, 2016.

Dobrowolny, Wolfgang. *Ashram in Poona: Bhagwan's Experiment.* New York City: DRB [1979] 1981.

Dunne, Anthony. *Hertzian Tales: An Investigation into the Critical and Aesthetic Potential of the Electronic Product as a Post-optimal Object.* London: Royal College of Art, 1997.

Dunne, Anthony, and Fiona Raby. *Design Noir: The Secret Life of Electronic Objects.* Basel: August/Birkhäuser, 2001.

Elden, Stuart. *Sloterdijk Now.* Edited by Stuart Elden, 1–17. Cambridge: Polity, 2012.

Haraway, Donna. *Staying with the Trouble.* Durham: Duke University Press, 2016.

Hexham, Irving, and Karla O. Poewe. *Understanding Cults and New Religions.* Michigan: Wm. B. Eerdmans, 1986.

hooks, bell. *Feminist Theory: From Margin to Center.* Cambridge, MA: South End Press, [1984] 2000.

Jain, Sarah S. Lochlann. *Malignant: How Cancer Becomes Us.* Berkeley: University of California Press, 2013.

Kember, Sarah. *Imedia: The Gendering of Objects, Environments and Smart Materials*. London: Palgrave Macmillan Pivot, 2015.

Koskinen, Ilpo, John Zimmerman, Thomas Binder, Johan Redström, and Stephan Wensveen. *Design Research Through Practice: From the Lab, Field, and Showroom*. Burlington, MA: Morgan Kaufmann (Elsevier), 2011.

Latour, Bruno. *Science in Action*. Milton Keynes: Open University Press, 1987.

Latour, Bruno, Pablo Jensen, Tommaso Venturini, Sébastian Grauwin, and Dominique Boullier. "The Whole Is Always Smaller Than Its Parts—A Digital Test of Gabriel Tardes' Monads." *The British Journal of Sociology* 63, no. 4 (2012): 590–615.

Linton, S. *Claiming Disability: Knowledge and Identity*. New York: New York University Press, 203, 1998.

Malpass, Matt. *Critical Design in Context: History, Theory, and Practices*. London: Bloomsbury, 2017.

Musil, Robert. *The Man Without Qualities*. Edited by Burton Pike with Sophie Wilkins. London: Picador, 1966.

Oliveira, Carlos, Peter Sloterdijk, and Carl Hanser. *Selbstversuch: Ein Gespräch Mit Carlos Oliveira*. München: Verlag, 1996.

Orr, Jackie. *Panic Diaries: Genealogy of Panic Disorder*. Durham: Duke University Press, 2006.

Poewe, Karla O. *My Apprenticeship: An Intellectual Journey*. Calgary: Vogelstein Press, 2018.

Raby, Fiona. "Interview." In *Darkitecture: Learning Architecture for the Twenty-First Century*, edited by Gerrard O'Carroll, 84–87. London: Two Little Boys, 2012.

Rajneesh, Bhagwan Shree. *Chapter Two: The Greatest Discovery There Is*. Osho Online Library. Accessed January 23, 2018. http://www.osho.com/iosho/library/read-book/online-library-sex-condemned-joke-8251a6bd-c62?p=84c4f2236f2b261789ef7a6d9dd842b7.

Roberts, Celia. *Messengers of Sex: Hormones, Biomedicine and Feminism*. Cambridge: Cambridge University Press, 2007.

Rosner, Daniela. *Critical Fabulation: Reworking the Methods and Margins of Design*. Cambridge: MIT Press, 2018.

Schön, Donald A. *The Reflective Practitioner: How Professionals Think in Action*. New York: Basic Books, 1983.

Sloterdijk, Peter. *Spheres, Bubbles: Microspherology*. Translated by Wieland Hoban. Vol. I. Los Angeles: Semiotext(e), 1998.

———. *Spheres, Globes: Macrospherology*. Translated by Wieland Hoban. Vol. II, Los Angeles: Semiotext(e), 1999.

——— *Spheres, Foams: Plural Spherology*. Translated by Wieland Hoban. Vol. III, Los Angeles: Semiotext(e), 2004.

Sloterdijk, Peter, and Gesa Mueller von der Haegen. "Instant Democracy: The Pneumatic Parliament(R)." In *Making Things Public: Atmospheres of Democracy*, edited by Bruno Latour and Peter Weibel, 952–58. Cambridge: MIT Press, 2005.

Stein, Gertrude. *The Autobiography of Alice B. Toklas*. London: Penguin Modern Classics, [1933] 2001.

CHAPTER 2

Situated Design Reading: The Air Pockets in *Spheres*

Abstract If philosophies of media can be read as a novel, where are designers when reading *Spheres* or *doing* spheres research? The answer is autobiographical so as to assess whether *Spheres* is an appropriate design philosophy. I remember what *Spheres* theory was staked upon: that's being-in the design of the depressive sphere. I bring together new and revised feminisms related to *Spheres* from design ethnographies and philosophy over a thirty-year period, beginning with a vignette about protesting women in 1968.

Keywords Protest · Rally · Reflect · Remember

REPEATING RAVE REVIEWS

Designers gathered in Falmouth University (Britain) at "Networks of Design" to hear Bruno Latour, their designers' philosopher. However, Latour steered the network towards his colleague, Peter Sloterdijk, whom he considered as a philosopher of design (2008). Latour insisted that it was because Sloterdijk takes seriously the *in*, of Martin Heidegger's "being-in" that designers should read *Spheres*. "Being-in" is also referred to as "Dasein", which for Sloterdijk, is a "suspendedness in nothingness" (ibid., 89).

© The Author(s) 2019
A. Windle, *A Companion of Feminisms for Digital Design and Spherology*,
https://doi.org/10.1007/978-3-030-02287-7_2

This seriousness about Dasein is what makes his philosophy so exciting for people like you who are bombarded with offers to redesign everything from chairs to climates. You cannot indulge anymore into the idea that there are, on the one hand, objective material constraints and, on the other, symbolic, human subjective ones.

—Latour (2008, 7)

Latour and Sloterdijk have both advocated for their separate theories of networks and *Spheres* to designers. They've exhibited together at ZKM (Zentrum für Kunst und Medientechnologie Karlsruhe) in "Making Things Public" (MTP) and presented to architecture students at Harvard University (TTM). Sloterdijk has also individually presented to architects at Delft University (INS). Since *Spheres*, Sloterdijk's papers, presentations and design exhibitions are loosely for "atmosphere designers", meaning those that design homemade bombs through to those whose designs benefit capitalism. The term found in *Spheres* (II–III) initially derives from *Terror from the Air* (TFA) because chunks of *Spheres* III were first published in TFA. Whether Sloterdijk's philosophical enquiry is more focused on design or social media, the rhetoric stays the same. It's exclusive, cynical and provocative.

Sloterdijk's philosophy rarely differentiates practitioners whether its architects, novelists, Latourian sociologists or priests (TTM), cartographers, hypnotists or performance artists (Latour 2009). These practitioners are brought together through an enquiry of space. Sloterdijk attends to bodies by hypotaxis—sentences about being-female that are hidden inside sentences about being-male. To not get lost in Sloterdijk's narrative maze as a designer, I gather a spherology of feminisms to make a point about where and how to read *Spheres*. It's a way of taking into account the situated body of a designer and I do so to protest against Sloterdijk, an author that made his name by philosophising on the activism of women.

Sloterdijk's rhetorical certainty partly derives from his attention to Neo-Platonist and theological writing styles. This style is derivative of the agora and the pulpit. Picture the "mansplaining" done while "manspreading" in the agora (Solnit 2015) in ancient mosaics that appear at the start of *Globes*, volume two of *Spheres* (II, 12). The term "manspreading" found its way into contemporary media as derived from American feminist writer Rebecca Solnit's notion of "mansplaining". I'm riffing here on the way philosophers talk about the bearded male philosopher in the agora (see Stengers 1997, 125; Law 2016, 21–22; and Sloterdijk III, 11). Sloterdijk pricks up the ears with performative speech as does Solnit's terms—mansplaining and manspreading—which are not altogether generative terms.

It can be argued that by muddling Sloterdijk's rhetoric with Platonic dialogues, I am making a false feminist argument. The Platonic dialogues aren't an obvious example of mansplaining because Platonists didn't set out to use rhetoric as preachers do. However, my point is not one of conflation (i.e. combining Platonic thought and theology). Sloterdijk in various forms of address (be that Platonic or Augustinian) and across history limits the ideas of women. Here's the muddle. Platonists rely on developing a voice through fictional and mythical characters. Furthermore, there were at least two women in Plato's academy—Axiothea of Phlius and Lasthenia of Mantinea. So, to argue against myself, perhaps there is no such thing as manspreading in the ancient agora that is neo-Platonic.[1] There is more to be said about rhetoric and feminisms, but this is where other theo-classical philosophers may want to interject.

Remembering Whose Cynical Reason?

Feminist philosopher Christine Battersby opens up an important debate about Sloterdijk's work prior to *Spheres*. An entire chapter of her book titled *The Phenomenal Women* (PW, 125–145) covers a vignette about a female protest. It counters a similar example by Sloterdijk appearing in the preface to his *Critique of Cynical Reason* (xxxv–xxxviii). Battersby's vignette recalls the 1968 student protests in Germany from when Battersby was herself a protester studying philosophy at Marburg-an-der-Lahn.[2] Protesting students in academic institutions disrupted classes across the country. In May 1968, a mass protest against the Notstandsgesetze (German Emergency Laws) took place. The laws for an internal state of emergency where basic constitutional rights were limited ensued and this affected education at all levels across the country. The German Socialist Student Union (SDS) spokesperson, Rudi Dutschke, had an attempt made on his life, leading to his death a decade later. Battersby only began to distance herself from her German activist friends in SDS when she discovered that fellow students had staged a protest in Munich on the anniversary of Kristallnacht: a violent attack on Jews and their property which took place across Germany between 9–10

November 1938. Munich had been the place of much Nazi violence and Kristallnacht marked the beginning of the Holocaust.

Sloterdijk is slightly younger, but yet a contemporary to Battersby. He may well have been starting his studies in Munich during 1969 and therefore could have also experienced the student protests in Frankfurt—the one he vignettes—whereby women flashed their breasts at W.G. Adorno.[3] Adorno was about to give a speech in a lecture hall at the Frankfurt School of Philosophy but was interrupted by female student protesters who took to the podium. Sloterdijk pointed out a parody in Adorno's actions when Adorno stood silently in opposition to the protestors. Adorno had been in support of the protest movement in the run-up to the Vietnam War, but his support wavered in 1969. Sloterdijk used this story to show how Adorno's negative dialectics are inadequate ways to disrupt the educational bourgeoisie of that time, but this is a partial account of one of many protests between 1968 and 1969—of bodies without words and action without situated meaning.

This vignette helps Sloterdijk to argue *against* Adorno. He points out that Adorno did not side with the student protest and stood silently aside. Yet, what these female protestors had to say is missing in Sloterdijk's account. Seemingly, only their embodied methods matter. They held a banner saying, "if Adorno is left in peace, capitalism will never cease", and threw petals onto Adorno's head (Müller-Doohm 2005, 475). Before and after the vignette, Sloterdijk's writing is scattered with references to the thrown flowers as "Pasolinian flowers of evil, and Freudian deadly nightshade" (CCR, xxxvi) and "zarre thought-flowers, red, blue and white, shimmering in the colours of beginning" (CCR, xxxviii). These are the same colours of the revolutionary flags of the Provisional Revolutionary Government of Vietnam (another issue of the '68 protest), as well as France, Chile and Cuba. These blooming allegories of protest start Sloterdijk's interest in the "deep idea of world annihilation" (CCR, xxxviii). It is through the protest of women that he begins to figure the depressive sphere before writing *Spheres*.

Battersby disagrees with Andreas Huyssen, a scholar of German and Comparative Literature at Columbia University, who wrote the foreword to CCR (ix–xxv). This is because Battersby cannot see how a feminist argument would strengthen Sloterdijk's ideas, as Huyssen suggests. I'd like to argue that remembering the detail of the lace trimmings as Battersby points out is crucial to understanding *Spheres* across generations of feminisms.

Battersby's vignette applies the writing of Adorno *against* Sloterdijk by drawing metaphorical meaning from the lace trimmings on the bras belonging to the student protestors (PW, 146). As Adorno says, "the eternal is more like the lace trimmings on a dress than like an idea" (Adorno in Battersby, 231). Battersby makes clear that this is a hetero-act of sexual violence by Sloterdijk *against* Adorno and attends to the detail that Sloterdijk removes (PW, 145).

Sloterdijk also debunks Adorno through Kynicism—a parody on Adorno's use of cynicism—to point out the falsehoods created by cynical prose. Battersby debunks Sloterdijk through Adorno's notion of trimmings (of which Adorno borrows from Walter Benjamin (PW, 145)). What is productive in Battersby's debunking is the awareness of the ways that Sloterdijk's theory was made through sexualised embodiment, typical of the explications found in *Spheres*. Sloterdijk vignettes women, their bodies and sexual liberation as the depressive sphere repeatedly in *Spheres*.

Rallying, Repeating and Remembering

Sloterdijk's account of female activists bearing breasts was used to make a philosophical stand *against* Adorno and the Frankfurt School of Philosophy. He wrote *Spheres* where continental philosophy was permitted in an art, design and media institution (ZKM). It's also where the ideas of Battersby still gain traction at the University of the Arts London, among others. It's not surprising to see that in the edited collection, *Sloterdijk Now* (SN), that Sloterdijk's vignette reappears.[4] This time, American philosopher, Babette Babich, writes both *with* and *against* Sloterdijk's ideas to highlight another female absence.[5]

I'll do some feminist connective work to suggest that Babich adds more lace trimmings (to use Adorno's metaphor) to Battersby's feminist philosophical account even though there is no direct referencing to one another. Babich's original contribution is to note the missing female Greek cynic that could have shaped Sloterdijk's cynical-kynicism, alternatively (CCR). Babich remembers the work of Hipparchia of Maroneia, a cynical female philosopher from around 300 BC (SN, 31–2). This challenges Sloterdijk's masculinised account of cynical philosophy. Babich's naming of Hipparchia of Maroneia also places the emergence of cynical thought in a blurring of the domestic (along with the public sphere) especially when Babich refers to the women who bore their breasts as

"ÜbermenschInnen" (superman-woman). Sloterdijk's provocative speech has been mistaken for fascism writ large (see Couture 2016, 78).

Few, if any feminists, would extend Sloterdijk's contributions to discussions on "breeding" and "genetics" because it echoes towards National Socialist eugenics (i.e. III, 700–720). However, Sloterdijk directly cites Friedrich Wilhelm Nietzsche's work when referencing "breeding". It isn't Nietzsche's ideas that are the issue because his work has been taken up by both left- and right-wing theorists. It is directly because of his sister, Elisabeth Alexandra Förster-Nietzsche, and Adolf Hitler's interpretation of Nietzsche's ideas that an anti-Semitic argument ensued. Martin Heidegger and Förster-Nietzsche were, however, both Nazis and while Nietzsche was not directly related to fascism and his ideas pre-date Nazism, his work, and that extended by his sister, is not feminist.

BEING-IN DESIGN ETHNOGRAPHY

To add another line of embroidered arcs around *Spheres*, I take inspiration from the French and British ethnographers, Sophie Houdart (2008) and Nina Wakeford (2011). Their ethnography of design practice and design ethnography, respectively, gives a feminist reader the resilience to persist with spherology for design because this secondary literature does just that. To my knowledge, Houdart and Wakeford published the first two papers which take Sloterdijk's metaphor of foam (think "co-isolation" for now, but more to come in chapter three), and do so as a theoretical enquiry into the social spaces of design and designing. Houdart's work considers the utopian *Spheres* that individual male architects and designers construct to display future visions at the Japanese World Fair. Whereas Wakeford advocates for more active ethnographic methods to counter the proliferation of sentiment analysis (performed with computational algorithms) taken up by industry practitioners at the Ethnographic Praxis in Industry Conference (EPIC), in Boulder, Colorado (2011). In my view, Wakeford's paper is a discussion of the proliferation of machine-learning techniques adapted for researching the designing of the home. While Houdart focuses on the creation of national spaces created in the design studio, Wakeford highlights another artifice: that of analysing human emotion in the home by machine-learning (i.e. sentiment analysis). There's no cynicism in either of their writing, but rather, accounts of where design intimately intervenes and where it creates technological visions of dwelling in the future.

Both ethnographers take up Sloterdijk's foam and bubble metaphors. They work only with what is helpful or necessary, i.e. the furthering of sociological thought about material practice. I'd argue, however, that anyone interested in *Spheres* should start with Battersby and Babich, before reading *Spheres* and proceed with Houdart and Wakeford's account of design. However, there's a few other writers to get to grips with too.

The translator, Luc Peters, adds an autobiographical note that's predominantly a liberating masculine gesture of creativity at the end of Sloterdijk's Delft presentation: "Besides working as a manager and lecturer, he indulges in music and beats the shit out of his drumkit" (INS, 251). Violent "speech acts" add to Latour's recommendation for designers to read *Spheres*. "Speech act" is a linguistic term used in philosophy by academics like the American queer-theorist Judith Butler (1997). Butler's work on violent speech acts has influenced contemporary thinking ranging from the Peace Movement in Palestine to inspiring Russian feminist activists like the Pussy Riot, imprisoned for their political punk protest songs *against* Vladimir Putin. Inspired by and working *with* Wakeford and Houdart's approaches to *Spheres*, next up is an autobiographical vignette about where designers may be situated when reading *Spheres*, encompassing both the domestic and public spaces of design practice.

READING *SPHERES* AS A DESIGNER

> The apartment-dweller is an individual who tries to marry himself or herself and to form the perfect couple with himself or herself.
> —Sloterdijk [Peters trans.] INS, 249

The statement above is just the sort of sentiment that excites Peters but aggravates feminists. It aggravates me because the statement assumes combined heterosexuality and narcissism of the apartment-dweller. It's this detail that Houdart and Wakeford leave behind, and it's their work that generatively moves a design reading of *Spheres* on. However, there's a bridging point to make about design practice that generatively stays with the trouble of the aforementioned quote. Practitioners in both art and design attend to the materiality of a text when reading in situ whether or not this reading is done adjacent to the making of creative work, be that in a study, studio, office, laboratory, library, lido, kitchen or café. It matters where and how a text is read and if indeed one is alone or in company, caring for others.

What follows are two recommendations for located reading that were given to me by the painter and lecturer, Roxy Walsh, when studying fine art (print) as an undergraduate student at Newcastle University.[6] Read Marcel Proust's first of seven volumes "Swann's Way" ([1913] 2000) in the bath, in England during the autumn. Bathing ends when fingers and toes start to wrinkle. Here's another: read Jacques Derrida's chapter "Différance" in his book entitled, *Margins of Philosophy* (1972), in a closed room with enough supplies to last in one sitting.[7] These texts differ in that the fiction is to be approached in a relaxed manner and the philosophy in a disciplined way. *Spheres* requires both methods.

READING *RATHER* THAN RUNNING

Where does Sloterdijk expect readers to be located when reading *Spheres*? In the ancient agora, in a studio, on the way to a world expo or in the homes where ethnographers hang out to design through experience? Reading in a suitable location can be understood from a range of concepts drawing on "situated knowledge" prevalent in feminist technoscience and STS, ranging from human-computer interaction through to sociology (i.e. Lave and Wenger 1992; Stengers 1997; Suchman 2007; Star 1994; Wakeford 2016), but it's also implicit across pedagogies of art, design and media communication (i.e. Orr et al. 2004, and Rughani 2013).

If one's carbon footprint, working practices and bank balance permit, my suggestion is to read the trilogy in the air, alone, at altitude, across hemispheres but still within the earth's atmosphere. I say this cynically because *Spheres* is aimed at the mobile designer. In recent years as a design researcher, I used a long-haul flight to read *Spheres*, jettisoning running trainers and fiction to make room—to read *Spheres* linearly from cover-to-cover. Accompanied by multicoloured post-it® notes to stick in the book or on the plastic backing of the airline seat, I cancelled out interruptions like pre-recorded inflight announcements and duty-free sales by listening to music on headphones. It's time to stop reading when turbulence makes for an uncomfortable read. For additional and intermittent readings of *Spheres*, I move to regular reading in between home and work, between apartment and agora by working agile between settings of low- and high-income studio practice.

Daily reading on the move means that design equipment like laptops, chargers, Sharpies® and post-it notes takes precedence for bag space when commuting to a university, conference or into-the-field. My copies of

Spheres rest on a bookshelf in my studio-office at work, or temporarily on a shelf at home in my small London rental. I live six minutes from work—my adjustments for living with an autoimmune condition. At home I free-cycled books to make storage space for *Spheres*. My point on the labour of reading is that it's a substantive annoyance to read the philosophy as it was intended. It's awkwardly heavy and is prodigious for space. Each book is full of spherical ideas, but its squareness doesn't match the foam proportions of capital cities with expensive housing. That's because Sloterdijk most likely wrote it in the industrial town of Karlsruhe, Germany, and perhaps when he lived in Pune, India. As a point of reflection, philosophical novels promoted to designers need to attend to the physical spaces that designers inhabit. Otherwise, they will jettison designing for design thinking or even space for the trappings of intimate life.

More Airings on *Spheres*

My first studio after graduating (before I joined a group of women at the Candid Arts Studios in London) was half an airing cupboard (where actual lace trimmings got aired). The result? Work was small-scale and thematically domestic. It therefore matters where and how *Spheres* is read and where we creatively practice. It matters whose design spaces are familiar and unfamiliar to the design reader of *Spheres* and whether philosophers support design for agoras or airing cupboards.

While reading *Spheres* as a novel as Sloterdijk advises—but doing so while travelling at speed thousands of metres in the air (like I suggest) and—vertiginously traversing through a millennium of philosophy (as Sloterdijk does), I recommend, while experiencing both, to feel the drag of gravity. Feel the air pressure that brings on cramps in the colon and/or uterus and pay close attention to the spaces for reading and how the body feels during this time. Reading isn't just the privy of academics in libraries; it's crucial to read in the places where practice takes place. Understanding how designers today make time and space for reading as intellectual labour will open up practical ways for writing-out dasein.

In conclusion, my protest *through* the written word is to advocate against a linear and direct reading of *Spheres* because politically it leads to beating the shit out of things (Peters) that incites hot-headed balaclava wearing (Pussy Riot). Machines will, I'm sure, pay intimate attention to the heat rising in the arcs of crochet balaclavas made at home, but to what end? As an original contribution to indifference? To be clear, my last two sentences are not intended to be gendered or binary, i.e. creative freedom against political oppression, to divide violent men from angry women.

Through my research into the preface (Battersby and Huyssen) and quasi-afterwords of *Spheres* (Babich, Houdart and Wakeford), this chapter was to help readers understand where *Spheres* loses touch with all kinds of feminist forebearers, be they female protestors without words or designers that work at home. In the next chapter, readers are guided to further air pockets in *Spheres* so as to attend closer to more literary Sturm und Drang about reading *Spheres* communally.

Notes

1. For agora style design studios see Myerson and Ross, *Spaces to Work*, 2006.
2. Insights taken from personal correspondence with Battersby, between December 2017 and January 2018.
3. Sloterdijk considers Adorno as a "crypto-Buddhist" that feels before he thinks, evident in his critical theory.

> It is the masculine world that it categorically rejects. It is inspired by an archaic *No* to the world of the fathers, legislators, and profiteers. [...] where memories of happiness are bound exclusively with a utopia of the feminine.
>
> —CCR, xxxiv–v.

   Sloterdijk hopes to differentiate from Adorno and other "literature of pain" (ibid., xxxvi–ii).
4. Christine Battersby and Stuart Elden were academics at Warwick University in the Departments of Politics and International Studies, and Philosophy, respectively. Perhaps Babich and Elden didn't cite Battersby in *SN*, because Elden left intermittently and edited while at Durham University, but it's a gap, that I and further design spherologists ought to remember.
5. Like Sloterdijk, Babich is a scholar that draws on Martin Heidegger and this might be why Battersby isn't cited by Babich (who draws on Adorno). When I asked Battersby why this might be, I received a lengthy reply. I'll recount only this:

> Philosophy is a brutal discipline for women — in addition, feminist philosophy and continental philosophy do, of necessity, marginalise one's impact, at least here in the UK.
>
> —Battersby, by email correspondence on the 13th December 2017.

6. Walsh's paintings of brightly coloured phalluses were exhibited locally in Newcastle as well as in New York during the 1990s at the time when Battersby was publishing *Phenomenal Women*. Both are, indeed, phenomenal women.
7. Better known is feminist Jeanette Winterson's recollection of smuggling Freud into her outside loo in Lancashire as a way to read beyond her religious indoctrination (1996, 153).

REFERENCES

Butler, Judith. *Excitable Speech: A Politics of the Performative*. New York: Routledge, 1997.

Couture, Jean-Pierre. *Sloterdijk*. Cambridge: Polity Press, 2016.

Derrida, Jacques. *Margins of Philosophy*. Translated by Alan Bass. *Différence*. Chicago: Chicago University Press, 1972.

Houdart, Sophie. "Copying, Cutting and Pasting Social Spheres: Computer Designers' Participation in Architectural Projects." *Science Technology Studies* 21, no. 1 (2008): 47–63.

Latour, Bruno. 2008. "A Cautious Prometheus? A Few Steps Toward a Philosophy of Design (With Special Attention to Peter Sloterdijk)." Accessed August 8, 2017. http://www.universal-publishers.com/book.php?method=ISBN&book=1599429063.

———. "Spheres and Networks: Two Ways to Reinterpret Globalization." *Harvard Design Magazine 30* (2009): 138–44.

Lave, Jean, and Etienne Wenger. *Situated Learning: Legitimate Peripheral Participation*. Cambridge: Cambridge University Press, 1992.

Law, John. "Modes of Knowing: Resources from the Baroque." In *Modes of Knowing*, edited by John Law and Evelyn Ruppert, 17–58. Mattering Press, 2016. https://www.matteringpress.org/.

Müller-Doohm, Stefan. *Adorno: A Biography*. Malden, MA: Polity Press, 2005.

Orr, Susan, Margo Blythman, and Joan Mullin. "Textual and Visual Interfaces in Design Education." *Art, Design & Communication in Higher Education 3*, no. 2 (2004): 75–80.

Proust, Marcel. *Search of Lost Time. Swann's Way*. London: Penguin, 1913.

Rughani, Pratap. *The Dance of Documentary Ethics*. Edited by Brian Winston, 98–109. British Film Institute, London: Palgrave Macmillan, 2013.

Solnit, Rebecca. *When Men Explain Things to Me*. London: Haymarket Books, 2015.

Star, Susan Leigh. "Misplaced Concretism and Concrete Situations: Feminism, Method and Information Technology." In *Boundary Objects and Beyond: Working with Leigh Star*, edited by Geoffrey C. Bowker, Stefan Timmermans, Adele Clarke, and Ellen Balka, 143–71. Cambridge: MIT Press, [1994] 2016.

Stengers, Isabelle. *Power and Invention: Situating Science*. Translated by P. Bains. Minneapolis: Minnesota Press, 1997.

Suchman, Lucy. *Human-Machine Configurations: Plans and Situated Actions*. Cambridge: Cambridge University Press, [1987] 2007.

Wakeford, Nina. 2011. "Replacing the Network Society with Social Foam: A Revolution for Corporate Ethnography." Accessed August 8, 2017. http://epicpeople.org/wp-content/uploads/2014/09/Wakeford_repla.pdf.

Wakeford, Nina. "Don't Go All the Way: Revisiting 'Misplaced Concretism'." In *Boundary Objects and Beyond: Working with Leigh Star*, edited by Geoffrey C. Bowker, Stefan Timmermans, Adele Clarke, and Ellen Balka, 69–84. Cambridge: MIT Press, 2016.

Winterson, Jeanette, *Art Objects: Essays on Ecstasy and Effrontery*. London: Vintage, 1996.

CHAPTER 3

An Atmosphere Ethics for Digital Transcription and Reading *Spheres* Communally

Abstract How, where and who reports back on communal activities in academia matters. The reader might expect there to be writing about the academic reading group's discussion, (of who said what) followed by a discussion about consensus and cooperation. Instead, the debate is circumnavigated so as to discuss the events' atmosphere ethics as related to *Spheres*. Atmosphere ethics is intertwined with a feminist ethics of ambiguity and situations. I give an account of participant consent and three schisms resulting in non-consent. The social, digital and technical schisms are tied to the spatial setting of where the roundtable took place at the Royal Society of the Arts (RSA), London, and then how it was captured in an audio recording and data processed at the point of digital transcription. The digital design-in is to suggest that consent is always a matter of continual re-consent.

Keywords Ambiguity · Atmosphere · Ethics · Reading · Re-consent Transcription

It's a sunny spring morning and I'm walking over the River Thames on my way to the Adelphi Terrace in London. Twelve of us (academics) are meeting at the Society for the Encouragement of Arts, Manufactures and Commerce, now generally known as the Royal Society of the Arts (RSA). We are meeting to share our views on selected reading by Karen Barad (2003), Peter Sloterdijk (2009) and Isabelle Stengers (2005).

© The Author(s) 2019
A. Windle, *A Companion of Feminisms for Digital Design and Spherology*,
https://doi.org/10.1007/978-3-030-02287-7_3

31

An obvious choice would have been to pair two short talks for design students—Sloterdijk's presentation, "Talking to Myself about the Poetics of Space" with Bruno Latour's "Spheres and Networks: Two Ways to Reinterpret Globalisation: A Lecture at Harvard University Graduate School of Design" (2009). I'd argue that juxtaposing Sloterdijk's writing with feminist philosophers of science and technology helps to update and shift the ethical register for investigating sharedness in relation to art and design practice.

At the event, we pour ourselves coffee, pitch-black out of long plastic canteens and take our seats around a circular, mole-skin grey table, centred in a square neoclassical room—oblong in height and squarely stunted in length and width with the light seeping in through the sash windows. Scottish architects (and brothers) John and Robert Adam, designed the building where this informal reading group entitled 'Informed Matters—Digital Media Materialities' takes place.

The event is part-funded by the Higher Education Funding Council for England and is part of a broader "Research Fortnight" of events for the University of the Arts London. The reading group brings together five early-career researchers, members of the Informed Matters Community of Practice and a few research centres, hubs and labs.[1] The group includes doctoral candidates, early-career researchers, fellows, industry-led researchers, lecturers and readers and our many practices cover, though not exclusively, architecture, design, fashion, film, fine art, journalism, philosophy, photography, policy making and sound art.[2] The roundtable was popular and could have been filled twice over, but it would be naïve to suggest that this is evidence that Sloterdijk is *the* designers' philosopher—the initial question of this book.

Locating the reading at the RSA was a way to move out of our regular institutional environments and into a new setting that could symbolically contain a myriad of academic and industry-led practices (from designing clothes to creating digital tools). For these philosophical readings we didn't choose an agora or don beards, we came dressed in jeans, trousers, shirts and skirts, with flip-flops and trainers.

The investigation of reading groups as social activities is important to research but it tends to be studied in pedagogy. On the other hand, STS is more concerned with the industry-focused mega-conferences of late, following Sloterdijk's call (and Latour's repeat) for more research into the understanding of sharedness in extra-curricular activities. I focus on a smaller roundtable event because it helps to understand discrete "foamed clubs" as an "ethics of atmosphere" (III, 606).

In *Spheres*, foam refers to human phenomena at varying scales like socialising at a mega-conference or mass-interminglings of microbes. However, when a human body becomes objectified then it becomes a different matter, like when Sloterdijk names women as architectural units (TTM, 3). This phrase was the impetus behind my main proposition for the roundtable: *how might Sloterdijk's work be reparatively questioned through a feminist enquiry?* Metaphorically, "foam" helps to describe the "co-isolated" and "co-fragile" aspects of the RSA reading group. His references to foam cities are helpful when large venue spaces are adjoined to the ethos of the New Babylonians, Fluxus Existentialists and the Situationists (III, 614–8). However, the metaphor needs updating in relation to new media technologies like diary blogging or transcribing using software because Sloterdijk doesn't address any media communication technologies in relation to foamed clubs. Instead, he mentions technology groupings like orthopedists and automotive shareholders who gather in conference venues.[3] *Spheres* stays within a socio-spatial summation of foamed clubs (III, 605–6).

I assess an "ethics of atmosphere" from *Spheres* through ethical feminisms, specifically to situations involving automated transcription and machine-learning. This shifts the human centre of Sloterdijk's being-into human-machine relations. There's an ethical difference in attempting to create a public communal record (i.e. podcast or text transcription) and an individual account (i.e. blog-post or diary-entry) of a reading group. I'll discuss the former as an account of participant consent and re-consent when attempting to create a communal record of shared reading through (a) the methods of audio-recording and (b) digital transcription.

Triangulating a Polysemic Reading of Spheres

To my knowledge, all three authors chosen for the reading group do not cite or recommend each other's work but they all reach similar communities in feminist technoscience and design STS. The shorter text by Sloterdijk was a presentation for architects and designers in his *Spheres* trilogy (TTM). I had hoped to question the adjacent texts with creative practitioners through the feminisms of Barad and Stengers.

I envisaged that the reading group would debate how women could possibly be architectural units. While the writing by Sloterdijk and Stengers has appeared side-by-side (MTP), the same cannot be said of Barad and Sloterdijk. Stengers and Barad are only situated together by secondary authors that draw the two together, notably Deboleena Roy, scholar of women's, gender and sexuality studies (2011). Essentially, the author whose ideas bridge all three texts is Stengers, which matters because it's an instance of a feminist writer creating accessible and open-ended texts that are capable of broad juxtapositions (unlike I–II).

While the popularity of Stengers' text reflects the interests of the reading group members, I infer that it's Stengers who is the practitioner's philosopher, not Peter Sloterdijk. The text by Stengers is more inclusively written and broader in its account of practice than Sloterdijk's narrowed focus on architecture, approached through phenomenological terminology.[4] Sloterdijk makes no apologies for how difficult the text is for a practitioner new to philosophy. He expects his readers to be part of a social group that both have the time and resources to read as widely and as prolifically as himself (III, 782 and 791). This expectation is one of the most dated methodical assumptions of *Spheres*. Furthermore, Sloterdijk suggests that *Spheres* 1–II defy citation and "are excluded from the quotable realm from the outset" (ibid., 807).

Feminist technoscience is almost entirely *excluded from the potential quotable realm* of the trilogy and that's why there is no relation to feminist ideas in Sloterdijk's abridged text. He builds "strong" connections to social and systems theory writers like Émile Durkheim, Karl Marx, Georg Simmel, Adam Smith, Gabriel Tarde and Bruno Latour than he does for feminisms (ibid., 235 and 456) through an entirely heteropatriarchal lineage, or *pater philosophorum* (II, 348).[5] It may go unnoticed that Sloterdijk has completed a cursory reading of feminist and gender studies writing in STS (and beyond). This evidence is only to be found within the endnotes of the third volume of *Spheres* where the work of Hannah Arendt (III, 876 n17), Donna Haraway (III, 844 n102), Julia Kristeva (III, 886 n147), Luce Irigaray (III, 842 n76) and Lynn Margulis (III, 833 n28) are referenced. This detail was impossible to cover when approaching the abridged version of *Spheres* at the roundtable. There is no arrangement, chapter or single sentence in 3000 pages whereby these writers connect as co-located feminist literature. They are structurally weakened to the endnotes and to parody the logical tensions of Sloterdijk's foam metaphor, they are the co-isolated and seemingly co-fragile margins of *Spheres*. The endnotes are "the great unthought and blocked-out element of [Sloterdijk's] sociological attention", to parody Sloterdijk's call for research on extra-curricular sharedness (ibid., 606). Therefore, the tensions created by the foamic metaphor in relation to feminist literature aren't just stretched, they are broken. Hinting at this in the reading group didn't help because I couldn't get to the details of the matter I was attempting to raise. There was too much enframing taken for granted in the group's dialogue and without a thorough knowledge of the trilogy, this point wasn't easy to reach.

In *Spheres*, Sloterdijk disagrees with Haraway in a single sentence about "a post-structuralist turn in biology" (ibid., 186). However, the endnote related to this sentence indicates a small window where Sloterdijk had read excerpts from the reader *The Feminist Theory and the Body* by Janet Price and Margrit Shildrick. Was he reflecting on his criticisms from feminist writers like Christine Battersby whose work also appears in this reader ([1999] 2008, 341–59)? Possibly indirectly. However, Sloterdijk's issue is with Haraway's chapter entitled "Biopolitics of Postmodern Bodies", which shows that neither Haraway nor Sloterdijk create stable bridges across eithers' work (ibid., 203–14). Feminism is the foamed literature of the trilogy, but it's neither the subject nor the target of Sloterdijk's jokes. To bring the two together is uncomfortable and requires reading around their literature. While Sloterdijk disagrees with Haraway, I'll reverse this gesture to infer that Haraway (if in dialogue) would disagree with the biomedical use of Sloterdijk's language and would sidestep his human-centred argument as Haraway does for Hannah Arendt (Haraway 2016, 127–30). Many of the binaries (dichotomies for Haraway, explications for Sloterdijk) that Haraway charts (Haraway, in Price and Shildrick [1999] 2008, 209) can be found in *Spheres* with exceptions shown with a strikethrough (III, 209) (Table 3.1).

Although this reading is situated in the notes at the end of *Spheres*— arguably the least read part of the book—this is the most important entry-point into reading *Spheres* for a feminist-design-reader; it is not via Sloterdijk's abridged text set for this reading group. In the next section, rather than consider what the group intended to discuss, I will address its atmosphere ethics.

Table 3.1 This table takes an excerpt from Haraway's chart of dichotomies (left side of the table) and then applies it where these juxtapositions appear in *Spheres* by volume and page number (right side of the table)

Haraway's chart of dichotomies		Appears in the corresponding volume of Sloterdijk's spheres trilogy	Page no.
Small group	Subsystem	I, III, III	76, 233, 393
Biological determinism	Systems constraints	I, II, III, III, III	122, 383, 199, 213, 420
Freud	Lacan	I,	463–473
Organic division of labour	Ergonomics, cybernetics	I	401
Microbiology, ~~tuberculosis~~	Immunology, ~~AIDS~~	II, III	383, 185, 310

Note The strikethrough emphasis is added by Windle

FROM CONSENT IN-THE-ROUND TO THE LOOPS OF RE-CONSENT

To scale into the foam worlds of ethical consent, I'd argue that up to a point, the audio device placed at the centre of the roundtable mediated the negotiation of group consent. It took five minutes to gain informal consent for the audio-recording but several months to coordinate re-consent, in order to turn the audio into a publicly shareable, text-based document. The performative placement of the device on the table was the beginning of an ethical discussion around individual participant consent for what aimed to be a communal record. The Sony™ IC Recorder (ICD-UX513F) is a discreet device but visible enough to make everyone aware of it beyond the moment of gaining group consent. Here is the recorded question of consent I asked:

> Is everybody okay if I record? Just because what you're saying is great — but is that alright if I do it; it's not for any hidden agenda other than to just try and capture what we've got and then disseminate it to everybody if you would like?

All participants were asked for their informed consent three times: once verbally (before the roundtable began), again in writing for the transcription and then, third, for use of the text in relation to the publication of this chapter (after the roundtable). A consent form was circulated for gaining re-consent. It was the above statement that re-consent hinged upon, which was emailed after the event. The consent form was approved by my university's ethical committee for a communal publication of a co-edited transcription (to be made public). Bear in mind that two very different public records already existed at this time, made by two participants of the group. Academic blog entries often do not need ethical consent because they are personal recollections akin to diary entries, but it's not that simple. It depends on the author's writing intentions, and that matters.

I intended for a communal document to open up academic discourse-in-the-making, considering that this would be insightful, particularly for the students who took part. This ethical attitude is positively associated with the freedoms of improvisation and "rudimentary" expression (Windle 2014). Additionally, a raw document doesn't need extensive editing. While publishing a raw account may, to some readers, seem "absurd", I saw it as an expression of ambiguity relating to a tradition of writing akin to the writings of Gertrude Stein. Stein's techniques were associated with Dadaism—most closely with Marcel Duchamp—rather than directly with Dada Ball or Dada Tzara. According to Marjorie Perloff, the American scholar of English, "[Stein] in part because of gender, belonged to none of the Paris cenacles so prominent in her day" (2002).[6] This impetus was countered by two participants who wanted an edited, perhaps even a "strong", public record. It could be inferred that the fragmentary discussion was also considered embarrassing or chaotic and not ready or suitable for public dissemination. My assertion to publish the rawness is also to pursue what feminist philosopher Simone de Beauvoir considers the freedom "to assert that its meaning is never fixed" (EA, 129). In order to create a "strong" editorial voice, I wasn't comfortable with editing the words of others for the sake of a clear document. I hoped to convey the sense of engagement between the reading group members and to show how debate emerges, interruptedly. The ethical issues that follow stem from the difficulties of reaching group consensus. This happened during the collection of individualised re-consents for the limited editing of a communal document. The writings of de Beauvoir are essential to reading *Spheres*, particularly when attempting to apply Sloterdijk's ethical ideas of atmosphere to digital design research. Perhaps my ethics form did not suggest enough of this methodological intent, but it certainly led to questioning what is a communal document and furthered my initial gender enquiry into *Spheres*—towards an understanding of rudimentary expression and the ethics of communal consent.

It's problematic to build on an ethics solely derived from Sloterdijk's account of atmosphere NOT because it is too instructional (it had copious ethical clauses) or too open-ended (ethical suggestions are vague or general), nor because his explications of atmosphere are intertwined with an ethics of defence—premising examples of gas, terror and war (III, 85–129)—or that the grand narratives of atmosphere refer to containments of glasshouses and globalisation (ibid., 46), or that he references an atmosphere of illness and lesions (ibid., 175), or hero-narratives of sacrifice and tragedy (I, 60; II, 91; III, 393), BUT because altogether this philosophy inspired by Jean-Paul Sartre, Friedrich Nietzsche and Martin Heidegger is so ethically positive about particularly negative *Spheres* of life that for an ethically inclined reader, what shouts from the pages is the principle of *mitigating risk*. How can such a horrific topic like gassing entire social groups in a labour supply-chain be so positively taken up as an aggressively dominant tactic of argumentative persuasion?

The writings of de Beauvoir, particularly *The Ethics of Ambiguity* (EA), are helpful in bringing to light how Sloterdijk's "atmosphere ethics" are troubling in their positive impetus for the negativity of terror. De Beauvoir considers the "abstract" qualities and "detached contemplation" of philosophical thought as an "aesthetic attitude" (ibid., 74). Even when de Beauvoir refers to the same destructive content of genocide and war, she does so without nihilistic intent (ibid.). Remaining in what Sloterdijk calls an "atmosphere ethics" requires a "situation ethics" (BOAB, 21), which means staying with the ambiguity found at the "margins" and in the "multiplicities of meaning" (Wakeford in BOAB, 82). Nina Wakeford draws on Susan Leigh Star's sociological account of "situation ethics", which in turn draws on de Beauvoir's "ethics of ambiguity", meaning that a researcher would not separate the socially good from the bad universals of moral ethics (de Beauvoir, EA; BOAB, 150 and 161; and Wakeford in BOAB, 76). To explain further, the *how* of writing about genocide cannot be separated out from the *why* (i.e. nihilistic intent).

There are three modes of ethics taking place that matter. There is the ethical route suggested in *Spheres* (atmosphere ethics) which I counter with the work of de Beauvoir (ethics of ambiguity) and there is the situation of the twelve-person meeting to discuss readings. The term "situated" draws on the literature of situated knowledge and practice. A situated approach helps to understand precise instances of action and agency (again from Lave and Wenger 1992; Star 1994; Stengers 1997 and Wakeford 2016), which is helpful for cutting through the phenomenological abstractions in *Spheres*. That's also not forgetting the human-nonhuman agency yet to be discussed in the transcription work. Ethics matters both within the texts about spatial theories of *Spheres* and in the spatial ethical arrangements of shared reading.

Social Ethical Schism (1)

Several social issues arose which caused ethical consent to break down. The labour required for each participant to edit their own authorial words was a request that didn't see much return, with only two participants emailing transcription edits. The reading of a raw transcription was for some, already too much work. The effort required to gain consensus and to pursue a collaborative method was both too much to ask of the group, and of myself. I eventually reluctantly called time on the consent for publication. I would argue that non-return of an ethics form is a discontinuation of consent and therefore a potential deferral of non-consent and a potential way of discontinuing consent. That is, however, if consent was secured in the first place (as it was in this case). If it hadn't been secured in the first place, then there would be no initial consent.

The benefits of going through an ethical process consisting of both a discussion and form are logical and methodical. This processual and cyclical approach helps to remain within an "ethics of ambiguity". Wakeford adds that studio-based researchers (including sociologists and STS practitioners as well as other forms of studio art, design and film practice) should make visible the work that is done around boundary objects and the gathering of communities of practice so as to "make visible, without entrapment — ambiguity, and even to have an ethics of ambiguity" (BOAB 76). It's been noted that further issues around collaborative methods of participation have proven difficult to coordinate and reach consensus in other forms of collaborative research practices like film-making (Breeze 2015, 15–16; Milne 2012; Pink 2004, 5; Rughani 2013), as well as news-based chat and online forums in digital media (Hine 2000, 23–24). These are important texts for academic and industry-led researchers interested in expanding their ethical reading, rather than *just* relying on internal review board (IRB) guidance (if afforded that support).[7] Had I suggested that we publish an audio transcription rather than a visual transcription, would consent have been easier to gain? Let me state for the record, that no one sent a form back to say that they didn't consent; on the contrary, only one participant didn't respond.

In summary, I'll suggest that the ability of a textual transcription over an audio transcription to represent the reality of a reading group and invite consensus is a fallible idea.[8] Take, for instance, Sloterdijk's short text for the reading group seminar which was a transcription of the spoken word. His nine-dimensional space is condensed to be a paragraph which is a condensation of 122 pages of labyrinthian text and often uncontextualised imagery (III, 340–462). There is a loss and a gain to be made in the translation of one medium (lengthy prose) to another (spoken word), and vice versa. The loss is the detail of ideas, but the gain is new audiences and readership.

Social Ethical Schism (2)

In brief, not all practitioners want to be recognised in a collectively organised group: "just because individuals are perceived or classified as a group does not mean they will act as a group, as this requires the practical and political work of organising and mobilising" (Lury and Wakeford 2012, 40). I'd argue that the consent of eleven individuals does NOT outweigh the non-return of one participant re-consent form. Furthermore, the reluctance of two participants to use ethics forms ought to be taken into consideration even if their consent was gained in a signed form, particularly when their signatures were submitted almost six months after the request. All of this contributed to stalling and could be perceived as a method of withholding rather than withdrawing consent. There may be more at play, like the coercive control of academic discourse. There's always more to understand from indecision, but let's not mistake indecision for what de Beauvoir defines as an "ethics of ambiguity", which is a more active approach to ethics rather than one that is passive-aggressive.

Did this group feel like their rights to academic freedom were being impinged upon by an ethics process supported by an IRB which helps the researcher to take informed consent requests through an ethical process (Denzin 2003, 254)? For some participants, yes. Were potential competitive dynamics in this reading group at play? Absolutely, and potentially I'd hedge because of unuttered disagreements in the heteropatriarchal vocabulary of a masculine phenomenology.[9] Was my desire for a communal output asserted within a context of writing and publishing this book? Yes. Did consent get interwoven into what might be considered "fair game for the researcher or as the property of all authors and not to be appropriated for academic purposes" (Hine 2000, 24)? No, because two journal blog s existed in public and yet, there's no publicly available transcription of this event.

There are remarkable differences in the two blog posts. One cites and names those who contributed, funded and organised the event, while the other writes about the joint texts, timestamped to before the round-table gathered. Pre-event writing is perhaps full of foresight and a sign of a keen blogger, but it's also a potential way to control the sequence of ideas. To reproduce the event in some a priori way, without quoting any other participant, de-situates the ideas of reading together from the event's situatedness, which was crucial in bringing together the shared reading and the group.

These are the decisions and questions that could be asked if a reader were to replicate and extend the reading group. If that's a possibility, then I'd suggest starting with "atmosphere ethics", but staying with a "situated ethics" by way of an "ethics of ambiguity". Read the cited work of de Beauvoir, Star and Wakeford together but with the last chapter of *Spheres*.

Social Ethical Schism (3)

The difference between a "Communities of Practice" (CoP) and a "Research Fortnight" event is another important social schism because some, but not all, practitioners consider themselves as researchers, meaning the experience and opinions relating to participant consent and the use of ethical forms varied considerably. The following points about membership in relation to academic sharedness need explaining because these distinctions may have been at odds. While the reading group participants are not all members of the RSA, the group were "something else besides" (III, 281). The reading group participants are in "company" because we are "already joined and related to one another elsewhere" (ibid., 282). This is perhaps a reflection of the specialist institution that is the University of the Arts London. Often referred to as a monotechnic in relation to the Research Excellence Framework (REF), it is unique because it consists almost exclusively of art, design, media and communication subjects but there are other areas like cosmetic science and pedagogy to name but two. The inclusion of this event during the University's Research Fortnight was the overarching timeframe and research funding was the only financial resource. In short, a Research Fortnight is a series of university events that gather over a two-week period to celebrate research practice: in this case, the practice of reading. Even though a CoP can lead to research outputs, it is not a criterion for such a group at UAL; however, it encouraged a possibility to "increase the involvement of strategic partners in developing and delivering our curriculum and research".[10] To push the scale of the assembly of this reading group even further, it is worth noting that a roundtable brings together other assemblages like hubs, labs and centres of which research academics are affiliated too, like me to the DigiLab.

From an STS context, "communities of practice" are more broadly understood to go beyond academia and the creative disciplines to include wider groups of membership within science and technology. Star draws on the human-computer-pedagogic work of Jean Lave and Etienne Wenger, whose own emphasis in their book *Situated Learning: Legitimate Peripheral Participation* is drawn from communities of practice around the conception of learning (1992). They consider how participants move from illegitimate peripheral participation to full membership in a community of practice. Star builds on their premise to consider a community of practice from a series of encounters with objects, and in this chapter, those objects are three main readings, one audio-recorder and an ethics form for participant re-consent (BOAB, 152). To build on this idea further, the relation of text to creative practice provided much of the energy in the reading group discussion.

TECHNICAL ETHICAL SCHISM (4)

In a second reading group in 2017, I chose to use an ethics form together with the placement of the audio-recorder on the table, again before the event got started. I hoped to include the two participants who did not want to fill in participant consent forms into the process of constructing one—to engage them in ethical debate before the event took place, yet this again did not reveal anything new. The 2017 reading group all consented, but the same two participants signed the form and disapproved of having to do so even though they had been invited to co-draft the form itself. They chose not to disclose this to the group. And, everyone signed. What then ensued was a quick-fired communal blog post made that night by the two participants which summarised my contributions in a way I completely disagreed with, and I said no at every point of editing with suggestions to include what I actually said instead via the audio-recording. In a sense, I felt coerced into a communal record that was not communal (i.e. did not involve all participants) and was a "strong" account of a debate. Although there was an expression in the consent form to write a piece together, it was a process of writing *against* rather than *with* a community of practice. Producing the account at such a quick rate was not part of the agreement as there was an intention to professionally transcribe the event as an audio-recording before committing to an account in writing. The methods of writing were changed, pushed and stretched beyond the initial ethical intent.

Researchers need to plan for a non-return of forms rather than a straightforward "yes or no" to participant consent, which connects to a participant's right to withdraw at any time. Sectioning the form into multiple and co-isolated aspects of consent (tick boxes for opting-out or into consent) is only one ethical challenge, but from experience on numerous projects I've found that a lengthier form helps to situate a conversation about ethics so that ethical enquiry is not just a form, but a discourse for shared learning. The participant consent form is the outcome of participant involvement, but not an end to the ethical debate. There are merits to a short participant consent form such as leaving more time for the content of the activity. However, what I would suggest from my early experiences of pithier one-page R&D consent forms is that these are too abbreviated and potentially conceal and shortcut the need for ethical discourse. Regular practice of participating in research helps build empathy with participants and is a way to advance ethical methods.

DIGITAL ETHICAL SCHISM (5)

Matters of consent become even more complex at the point of digital transcription because a further constellation of nonhuman agencies arises. This section is inspired by the sociological writing of Les Back and his audio-recorder (Back in Lury and Wakeford 2012, 245). Like Back, I do have a strong connection to my digital audio-recorder, but unlike Andy Warhol, there is no anthropomorphic connection to be made to this form of media. When Sloterdijk reiterates Warhol's naming of his tape recorder as his wife (I, 401), he's repeating subservient relations of machines to humans, as well as to gender. This may be a neo-Platonic technique of philosophy, but it's still possessively sexist to read.

What I offer instead is a reading of the entangled agencies involved in transcribing to show how the notion of consent echoes beyond the voices staged in a room.

A digital audio-recorder helps to transcribe the audible word, enabling researchers time to take notes to self, saving ideas for another moment that may not be for sharing but for thinking over, when on one's own. An additional benefit of staying close to the transcription is that it's a record of attribution and listening, of incomplete speech, silences, over-talking and of how debate emerges over time.

In 2017, when I transcribed the reading group myself, I used a software called Dragon Nuance™ 5.0. Nonhuman agency intervenes and extends human consent when digitised transcription links to this online service. I use the same automated software that is used when I commission it as a professional transcription. Whether it is the researcher (in the 2017 reading group) or a professional transcribing (in 2016), this method begins with automation. The text is compared to the audio-recording manually by listening intently to the errors and gaps in the digital dictation. Transcribing group voices is tricky and more expensive, requiring more human listening because accents, volumes and speeds of intonation differ in group communication. When transcribing myself, I physically talk back into the dictation software while listening to the original transcription on headphones, therefore re-recording the discussion in real-time. The dictation software easily recognises my single voice input. There are other ways of transcribing multiple voices like enabling the software package to "learn" each voice in turn. Dragon Nuance™ 5.0 has extensive machine-learning capability, requiring the programme to communicate with an online voice recognition database. In 2017, it's usual for most dictation and speech software, like that included on every Apple Mac™ computer or embedded within Google Chrome™.

When transcribing voices through the assistance of dictation software, the notion of consent gets caught up in the intellectual property rights (IP) that a purchaser signs over to the dictation company. In the license agreement for Dragon™ 5.0, there is a paradox; the content assembled requires consent only from the person who owns the license. This may not be the person initiating the act of automated transcription (i.e. a commissioned transcriber). Here lies an overlap between current IP and ethics agreements. It means that researchers who transcribe in this way remain within an ethical ambiguity of a "politics of simultaneity and contradiction" (BOAB, 163).

For instance, after purchase, when downloading the application software to a computer it requires the user (for want of a better word) to do so over the internet—which is arranged by a separate company called "Digital River Ireland Ltd"—otherwise referred to in the privacy agreement as Digital River[TM].[11] In the privacy agreement (which is operating out of Ireland and in relation to Irish law) the transfer of personal information is made, in this case, to my purchasing details. Digital River collects other information called "tags" or "persistent cookies" and "web beacons" (also referred to as "tracking pixels", "1×1 gifs", "single-pixel gifs", "pixel tags" or "action tags") which are graphic images, usually no larger than 1×1 pixel, placed at various locations on this website. This commences from the point of download rather than the point of sale, yet for how long these tracking things still operate is undefined because this material is shared with other companies and services and the privacy statement does not cover data-processing by these third-party service providers.

From the point of using the application, the software itself operates with different permissions and is incorporated by an MIT[TM] License which is an approved open source initiative (2009–2013). Of mentionable note: first, this includes the software that connects to the "Dragon Natural Speaking Learning Center" where voice commands can be learned.[12] Second, the software also includes various language model optimisers, all of which are discussed in the 295-page Dragon Installation and User Guide (240–42).[13] Third, personal identifiers (addresses and names) can be added to the programme from email programs like Microsoft[TM] Outlook[TM] and Windows Live[TM], as well as Lotus Notes[TM], Gmail[TM] and other web-based email (ibid., 255). And finally, from Dragon Nuance[TM]'s website a researcher can learn about all the ways the collected voice-based interactions are used as datasets for further R&D research (of Natural Language Processing and Artificial Intelligence) by 140 researchers in labs throughout the United States, Canada and parts of Europe.[14]

What can be learned from the privacy agreements, intellectual property clauses and open source licenses of public voice data? Transcribing a reading group through an automated service permits the dictation company to keep a record of voice data and therefore the content too. Included in my 2017 software PoS agreement is the ability for the constellation of R&D labs to use this data and content unrestricted for their future innovations. I can infer that participant consent would be better termed as "perpetual re-consent" meaning occurring repeatedly because there are several processual moments where consent can reoccur. However, on account of the open, temporally future-extensive ambiguity of automated and distributed service networks, what is more complicated is the potential for "perpetual re-consent" to be legally defined as never-ending and always in flux. However, new data laws as well as debates about net neutrality may change this in 2018 and beyond.

Could aggregated data content be reversed in the future if the software company were to innovate on the content they collect? Yes, because there is no fixed temporality to the consent of data transfer after the PoS. There's also no *right to withdraw* data from this service. The careful work done regarding consent by an ethical researcher may be undone by future R&D inventions, not yet invented at the time of using Dragon Dictate. Informed consent is up-turned by lengthy statements in large print distributed into different aspects of the products' user-journey of service and use, rather than hidden in the small print otherwise known as terms and conditions.

The workaround is to switch-off the ways the software on the computer communicates (online) with Dragon Nuance™'s machine-learning hardware. The network connection can be monitored to check that this is the case by using the application, "Little Snitch", a tool more readily used to check what's reducing the speed of an internet connection when online (like adverts). Design researchers should be aware—through a social understanding of science and technology—of the various ways in which machine-learning capabilities extend the technicity of human consent into the intellectual property domain of digital media and global commerce. If dictation companies no longer give an option for consent used by third parties then the ethics of consent is wrapped up a priori of using a product. There's no ethical ambiguity. It shortcuts future ethical debate about how dictation companies may use the content going forward. For more on machine-learning, see Chapter 4, and for legal modes of knowing, Chapter 5.

To return to my proposition from the reading group, I still wonder what Sloterdijk meant when describing "women as architectural units" (TTM 3)? To be clear, Sloterdijk is not inferring that men and women are a statistical unit of data or as a feminised interior, but rather to the apartment as a "studio of self-relationships" and "self-couplings" (III, 547). However, in the context of the breakdown of consent, all humans are merely units of digital "voice" data and transaction information.

Retrospection

Aiming to organise a convivial space is just one aspect of an ethics of atmosphere. What happened beyond that initial cooperation remains unreported as a communal document, but that isn't to suggest sharedness didn't take place. I wouldn't suggest, to parody Sloterdijk's abridged text, of "talking to oneself", because it's an inward-looking authoritativeness that gives voice to grand narratives. Sloterdijk's register of "talking to oneself" could be seen as a warning against too much self-recognition as narcissism and/or depression.

What collaboration took place in the Shipley room was in the coming together of an insulated "society foamed in [professional] clubs" (III, 605). To make practical an atmosphere ethics of foamed clubs, a situated ethics of ambiguity is necessary. The ethics of this paper (extended through automated transcription) became entangled with the ethical implications of nonhuman agencies which stretches the limitations of human ethical consent. Re-consent has reached the domain of machine-learning, which disrupts the human stability of being-in in *Spheres*, as the only mode of ethical knowing.

Gaining "100 percent" ethical consent before participation is another fallacy of successful ethics practice. A researcher may obtain relief from this simple ethical arrangement but there is relief to be gained from disagreement and sustained ethical discourse. When considering consent as a lengthier *diplomatic* practice, a researcher may approach ethical consent less as a singular moment initiated with a form (like the up-front ethics approach), and more as an ongoing process of collaboration. In this case, the frequency of the engagement, the nature and the context of the research may require many points of additional consent. A researcher may even go as far as to say that upfront ethical consent is meaningless if not followed up. I'd conclude that ethical consent could be better understood or even termed as participant re-consent because an ethical issue may extend beyond but still relate to the initial consenting moment. Furthermore, it means a researcher has planned for the *right to withdraw* process right through to the method of employing a digitised and automated transcription service. Reading groups are a feminist issue of "reporting-back" and it's a way to help understand biotechnological progress in relation to ethics, machine-learning and educational commerce.

Excursus

Sloterdijk writes generous chapter-length excursuses in *Spheres*. I will permit only one, which is less to do with the structure of *Spheres* and more to do with keeping the discussion of the readings distinct from the ethics of atmosphere and making sure the intersectional issues of Sloterdijk's paper are the final words of this chapter.

Design STS should be concerned with the way academics are taught about women in design research and how to reflect on practice together. Sloterdijk's abridged text matters because it depends which design students see themselves reflected in the cases of design and if there is a self to recognise. It ethically matters if we figure tape recorders as wives into our readings of shared situatedness.

Sloterdijk's short paper was presented to designers at Harvard University (2009). The ideas of the reflective practitioner for designers is well-known, since Donald Schön's main text, *The Reflective Practitioner* (1983), is still on many academic reading lists. It is Schön's text that derives from the same school of design that Sloterdijk presents his paper.[15] The problem with bringing *Spheres* to the same place where Schön wrote this textbook is that this supports a particular lineage of design educational texts which requires ethical reflection. Sloterdijk argues that "reflexivity, and pampering belong inseparably together" as aspects of a luxuriant world and that "pampering" and other words that are considered "self-descriptions" are concealed in the use of words like "recognition" (III, 755). Here there is a departure from Sloterdijk's criticisms of reflexive writing and Schön's advocacy for a designer's reflective practice and pedagogy. This is not the community of practice-based literature a design researcher should readily accept, at least not without feminist advocacy, bridging works and a thorough understanding of feminist ethics.

Schön's text draws on a review of architectural education directed by Dean William Porter of the Harvard Graduate School of Architecture and Planning and Dean Maurice Kilbridge of the Harvard Graduate School of Design in the 1970s. The example is derived from Roger Simmonds, Donald Schön's graduate student. The case from Chapter 3 involves a "studio master" named Quist teaching Petra about a "screwy site" (1983). This gendered relation would have been a novelty at this time and would even be considered progressive.

Today, I notice the isolation of Petra within this architectural pedagogy in the professional design studio. I notice how Scandinavian names (chosen by Simmonds to protect anonymity) keep intact a mastership of architectural whiteness. According to Kathryn H. Anthony in her book, *Networks of Power*, the AIA (American Institute of American Architects) "began collecting data on race and gender of its members in 1983" (2001). From 1983, this is defined as "African-American" (and in the 1990s as "black American"), "Asian" and "Hispanic" men and women (Anthony 2001, 97–100). According to Bob Borson in 1998 (working on the same archive as Anthony), the number of licensed architects was 2100 (about 4% of all licensed architects were female and of any race) rising to 15.5% in 1991 and again to 20% in 2003 (Borson 2011).[16] In 1972 MIT had only 28 female students altogether and it wasn't until 1984 that Anne F. Friedlaender, the first female Dean was appointed to MIT (2003). The following year, in 1985, the first African-American woman was a licensed architect and member of the AIA. At the time of Simmond's study, the AIA had formed its task force on women and at the same time had created another task force on minority groups (Anthony 2001, 97). Equal civil rights for gay and lesbian designers did not appear until 1995 and a joint Diversity Task force was set up in 1992 (ibid., 105–6). In 1998 the National Diversity Conference inclusively brought together a range of designers, not just licensed practitioners but students studying engineering, architecture and other spatial design studies (ibid., 107). Similarly, drawn from my own archival research from the journal *Science, Technology and Human Values* (ST&HV), the first full report related to STS about a feminist issue was written about the 0.5% take-up of women in North American engineering by Sharon Friedman, who was, at that time Assistant Professor of Journalism at Lehigh University Bethlehem, Pennsylvania (1977).

Take a moment to pause and ponder that institutional network because it is another disciplinary reflection on institutional sharedness. Without the work of feminist critique, Sloterdijk would merely add to a North-Easterly American architectural legacy of white heteropatriarchal design literature. This adds to why reading *Spheres* intersectionally together with feminist literature is a counterpoint to the heteropatriarchal containment of being-in *Spheres* and its abridged text, performed at Harvard in 2009.

Notes

1. Betti Marenkoand I supported each other in the organisation of our UAL research fortnight events and while I suggested the authors, Marenko chose the specific Stengers' text.
2. An Early Career Researcher as defined by the Arts and Humanities Research Council is "an individual who is within eight years of the award of their PhD or equivalent professional training, or an individual who is within six years of their first academic appointment". Last accessed 7/4/17 from: http://www.ahrc.ac.uk/skills/earlycareerresearchers/definitionofeligibility/.
3. Employees of IBM^{TM} and BMW^{TM} visited Rajneeshpuram (2018).
4. Here's a brief list: bewegheit–movedness (I, 333), existenz–existence (ibid., 628), kosmos–everywhere (II, 33), mitsein–being-with (ibid., 325), continens–ring (ibid., 402) and wohnung–dwelling/in-dwelling (III, 135–469).
5. Throughout *Spheres*, the adjective "strong" means "strong relationship" putting humans centre-stage of Sloterdijk's social concepts, contrary to STS (III, 214).
6. Simon Willmoth pointed out the dis-connection of Dada groups to Stein. It was this observation that led me to find the work of Perloff.
7. I was a college representative and member of my University's Ethics Committee in 2017.
8. Having the audio-recording as the main output rather than the text wasn't an option on the ethics form.
9. See Sara Ahmed's book on *Queer Phenomenology* (2006).
10. The publicly accessible criteria for UAL's CoP can be found here. Last accessed, 28/3/17 from: https://process.arts.ac.uk/content/ual-communities-practice-fund. All other paperwork is situated behind a password-protected university intranet.
11. Privacy Agreement. Last accessed 8/8/17 from: https://shop.nuance.co.uk/DRHM/store?Action=DisplayDRPrivacyPolicyPage&SiteID=defaults&Locale=en_GB&ThemeID=22100&Env=BASE&eCommerceProvider=Digital%20River%20Ireland%20Ltd.
12. Dragon Nuance^{TM} 5.0 Statement. Last accessed 7/4/17 from: http://www.nuance.co.uk/support/dragon-naturallyspeaking/index.htm.
13. Dragon Nuance^{TM} 5.0 Installation and User Guide. Last accessed 28/7/17 from: http://www.nuance.co.uk/ucmprod/groups/corporate/@web-enus/documents/collateral/dns13_userguide.pdf#page240.
14. Dragon Nuance^{TM} 5.0 foamed clubs of R&D. Last accessed 28/7/17 from: http://research.nuance.com.

15. The copy in my university library was taken out on average 4 times annually between 2005–2011. The subsequent digitisation of loans no longer enables a borrower to work this out at a glance.
16. In 2011, Bob Borson collated statistics from: The Beverly Willis Architecture Foundation, The National Center for Education Statistics: Degrees in architecture and related services conferred by degree-granting institutions, by level of degree and sex of student: Selected years, 1949–1950 through 2008–2009; and from the University of Illinois at Urbana-Champaign with additional statistics courtesy of the Almanac of Architecture and Design, Washington: Greenway Communications, 2004.

REFERENCES

Ahmed, Sara. *Queer Phenomenology: Orientations, Objects, Others.* Durham: Duke University Press, 2006.

Anthony, Kathryn H. *Designing for Diversity: Gender, Race, and Ethnicity in the Architectural Profession.* Minneapolis: University of Illinois Press, 2001.

Barad, Karen. "Posthumanist Performativity: Toward an Understanding of How Matter Comes to Matter." *Signs: Journal of Women in Culture and Society* 28, no. 3 (2003): 801–31.

Battersby, Christine. "Her Body/Her Boundaries." In *Feminist Theory and the Body: A Reader*, edited by Janet Price and Margrit Shildrick, 341–59. Oxon: Routledge, 2008.

Borson, Bob. 2011. "Life of an Architect." Accessed April 7, 2017. http://www.lifeofanarchitect.com/women-in-architecture/.

Breeze, Maddie. *Seriousness and Women's Roller Derby: Gender, Organisation, and Ambivalence.* Edinburgh: Palgrave Macmillan, 2015.

Denzin, Patrick. *Performance Ethnography: Critical Pedagogy and the Politics of Culture.* London: Sage, 2003.

Friedman, Sharon M. "Women in Engineering: Influential Factors for Career Choice." *Science Technology and Human Values* 2, no. 3 (1977): 14–16. http://journals.sagepub.com/home/sthOnline.

Haraway, Donna. "The Biopolitics of Postmodern Bodies: Determinations of Self in Immune System Discourse." In *Feminist Theory and the Body: A Reader*, edited by Janet Price and Margrit Shildrick, 203–14. New York: Routledge, [1999] 2008.

Haraway, Donna. *Staying with the Trouble.* Durham: Duke University Press, 2016.

Hine, Christine. *Virtual Ethnography.* London: Sage, 2000.

Latour, Bruno. "Spheres and Networks: Two Ways to Reinterpret Globalization." *Harvard Design Magazine* 30 (2009): 138–44.

Lave, Jean, and Etienne Wenger. *Situated Learning: Legitimate Peripheral Participation*. Cambridge: Cambridge University Press, 1992.

Lury, Celia, and Nina Wakeford. *Inventive Methods: The Happening of the Social*. London: Routledge, 2012.

Milne, Elisabeth-Jane. "Saying 'No' to Participatory Video: Unravelling the Complexities of (Non)Participation." Edited by C. De Lange, E. J. Milne, and N. Mitchell, 257–68. Lanham: AltaMira, 2012.

Perloff, Marjorie. 2002. "Dada Without Duchamp / Duchamp Without Dada: Avant-garde Tradition and the Individual Talent." *Electronic Poetry Center*. Accessed December 8, 2017. http://epc.buffalo.edu/authors/perloff/dada.html.

Pink, Sarah, László Kürti, and Ana Isabel Afonso. *Working Images: Visual Representation in Ethnography*. New York: Routledge, 2004.

Roy, Deboleena. "Feminist Approaches to the Enquiry in the Natural Sciences: Practices in the Lab." In *Handbook of Feminist Research: Theory and Praxis*, edited by Sharlene Nagy Hesse-Biber, 312–29. London: Sage, 2011.

Rughani, Pratap. *The Dance of Documentary Ethics*. Edited by Brian Winston, 98–109. British Film Institute, London: Palgrave Macmillan, 2013.

Schön, Donald A. *The Reflective Practitioner: How Professionals Think in Action*. New York: Basic Books, 1983.

Sloterdijk, Peter. "Spheres Theory: Talking to Myself About the Poetics of Space." *Harvard Design Magazine* 30 (2009): 1–8.

Star, Susan Leigh. "Misplaced Concretism and Concrete Situations: Feminism, Method and Information Technology." In *Boundary Objects and Beyond: Working with Leigh Star*, edited by Geoffrey C. Bowker, Stefan Timmermans, Adele Clarke, and Ellen Balka, 143–71. Cambridge: MIT Press, [1994] 2016.

Stengers, Isabelle. *Power and Invention: Situating Science*. Translated by P. Bains. Minneapolis: Minnesota Press, 1997.

———. "Introductory Notes on an Ecology of Practices." *Cultural Studies Review* 11, no. 1 (2005): 183–96.

Wakeford, Nina. "Don't Go All the Way: Revisiting 'Misplaced Concretism'." In *Boundary Objects and Beyond: Working with Leigh Star*, edited by Geoffrey C. Bowker, Stefan Timmermans, Adele Clarke, and Ellen Balka, 69–84. Cambridge: MIT Press, 2016.

Windle, A. "Automation and Design for Prevention: Fictional Accounts of Misanthropic Agency from the Elevator (Lift) to the Sexbot (Chatbot)." *Technoetic Arts* 12, no. 1 (2014): 91–106.

CHAPTER 4

By Count and Diagram, Is the Indexicality of *Spheres* Diverse Enough?

Abstract A book's indexicality (citations, bibliography, endnotes and references) matters because this is how ideas travel and authors become world-leading. Approached initially through counting, I evidence how Sloterdijk is primarily negligible through indexicality towards those who identify as female. What follows is a diagrammatic account of distortions in the *Spheres* index. This chapter's design-in is diagrammatic, using the tropes of *Spheres*. I argue that visualising the disciplinary of *Spheres* should be done by choosing the Voronoi over the Venn diagram because of the prevalence of casual sexism implied recently to the Venn. By cautioning the optimism to create new digital tools for searching indexes, I contemplate how new data visualisations may reinforce, rather than reveal, the distortions present in *Spheres*.

Keywords Citation distortion · Indexicality · Matilda Effect Venn diagram · Voronoi diagram

When I asked two female academics of photography and design to use Sloterdijk for a potential reading group (that of Chapter 3), they both politely declined with a similar declaration: "he's of a generation". When I replied that I would like to read *Spheres* to speak out on the troubles I have with Sloterdijk, there was a change of tone and the conversations became animated again. If Sloterdijk's trilogy could (for a moment) be considered as a meteorite, then this chapter is the result of the impact

© The Author(s) 2019
A. Windle, *A Companion of Feminisms for Digital Design and Spherology*,
https://doi.org/10.1007/978-3-030-02287-7_4

53

of the trilogy on its cited authors. How feminist writers are introduced matters, and I'm well aware that it's not usually through texts that bring multiple feminisms together.

Sloterdijk is an exemplar of a world-leading name whose references matter just that bit more because they set the trends for multiple disciplines. Meanwhile, everyday academics are revising their academic reading in relation to the quality/diversity nexus. Academics across the world have to make a choice. Academics either pick up the pieces that the trailblazing superstar left behind and in the process of examining will feel the impact of the meteorite from the shiny light. Or, they choose to ignore it and hope that very quickly the light dims as other more interesting work appears elsewhere. I've opted to do both. Here's a short reason why.

Sloterdijk is an exemplar of the thousands of academics and practitioners who don't cite women. So what? The experience of researching naming conventions gets complicated in the many other ways that words can injure. Here's how it does that for me. I've experienced threats—actual violence with time spent in a women's refuge—so the simple act of reading all the missing women in *Spheres* is a massive re-enactment of so much more. Now my re-traumatising may seem overblown to those who do not share my experiences, but partly it's why I do this work. Re-indexing *Spheres* requires the researcher to feel its effects when others would go in search of more interstellar reads. What's at stake in reading and writing about *Spheres* is more than just time because what we choose to read will shape our mood and thoughts within our bodies and will continue to shape our social performances within our everyday lives. Even though Sloterdijk deconstructs many of the spherical tropes that have sustained conceptions of individualism and difference, togetherness and belonging, awkwardness and isolation, I wonder throughout this chapter "who does Sloterdijk chose to include in *Spheres?*"

I advocate for a minimal 30% rule for citing female writers. It's not radical, but it's still activism enough for writing a companion text derived from the feminist literature of *Spheres*. In *Living a Feminist Life* Sara Ahmed chooses to cite no white men, so as to contribute to an "intellectual genealogy of feminism and racism" (2017, 15). This is important academic-activism from an author who has written and engaged in academic diversity work for many years (but has subsequently left full-time academia). Ahmed's approach would rule out writing a feminist companion like this one because the white-cis-hetero-patriarchy is a constraining starting point. Ahmed makes more room to write in her account of *Queer*

Phenomenology as a lesbian-writer-woman-of-colour and colour-full-lesbian-woman-writer (2006). I'd argue that *Spheres* readers do need to be reminded of why it matters to be fair to all women, men and those beyond binaries in relation to their age, colour, gender and sexuality. There are nuances too. So when I cite Adorno in Chapter 2 I'm aware of the forenaming, which Adorno reduced to "W.G.", a gender-neutral initialling. And, when I cite Colson Whitehead in Chapter 5, I do so knowing that he characterises Lila Mae, a woman of colour and a first-generation academic living in North America from the South. That's why tracing indexicality in relation to the main narrative remains critical. I show how obscure references in *Spheres* can interrupt a spherology that falls short of addressing a crisis in gender and an account of human pain.

Even with the constraint of writing a companion to a white-cis-heteropatriarchal author, I was able to double the citation rate to 60% by finding the obscured or absented women. This means that building on *Spheres* through its very small number of female references alone can create an entirely different togetherness, despite its initial anchor. Investigating the female references led to both missing and obscured women and men. Ahmed has done her own counts but predominantly via social media (i.e. Facebook™ and Twitter™). Conversationally, Ahmed learned that the counter-argument to cite non-diversely is often defended through passivity as an inert form of sexism (2017, 150). My less radical approach to indexicality, even though it is anchored to *Spheres*, still deals with the same issues that Ahmed raises to tackle the "why" of indexical obscurity. In terms of citing female scholars of colour—my companion concerns about 15% of all women I cite and that's why Ahmed's citation policy makes sense and goes further than me staying with the trouble of *Spheres*. However, at times in an academic career we are required to anchor to something other than a "good-mother" and so this companion responds to the irrelevant-single-authored-women-in-academia (Klein 1929, 434).

Indexes or rather indexicality (citations, references, quotations and indexes) is an assemblage of a unique intellectual community and the passing on of ideas from one scholar to another. To miss people out is one of the most basic forms of academic neglect, if not done accidentally.[1] I'm also conscious of the addictive kick of doing this work; the point is not to just look at all the forgotten women and leave it there; it means unpacking the explications (logics of production) inherent in indexicality. If indexing is done by someone other than the author, then the striking lack of diversity they have created may go entirely unnoticed.

While Ahmed moves into social media, I steer towards data visualisation to show how design methods can extend techniques of sexism and racism. I concentrate on the Voronoi (cellular shapes situated next to one another that do not overlap), an obscured diagram to be found in *Spheres*, rather than a Venn diagram (which intersects two or more overlapping circles). This comes at the end of the chapter after making assessments about *Spheres* and its indexicality. I do this to include diagrammatic explications of indexicality. I argue, rather than merely illustrate with the bubble and foam geometries of *Spheres*, to make a point about parody and to reveal just what it's like to *be-in-a-diagram*.

To Foam-Up the *Spheres* Index

Indexes to books are places for feminist intervention, innovation and invention. At a glance, an extensive index can help the reader to trace people, things and topics equivocally, but an index to a book is of course more than the sum of its content. The index to *Spheres* helps readers to refer to images and text together, but it's an authorial index which doesn't summarise ideas, subject or geography. Having no subject index would have made sense if Sloterdijk's aim was to debunk grand theories and destabilise an abstract understanding of being-in-the-world, but instead he sustains an interpretation of grand theory and narratives with renewed literary metaphors and tropes. I've not been able to find out whether (a) Sloterdijk commissioned an indexer via the publisher, (b) hired a doctoral/postdoctoral scholar for this work or (c) created it himself with the aid of digital software. However, the author is ultimately responsible.

Illuminating on the gaps in Sloterdijk's index has been a way of coming to terms with the (predominantly) male academics and designers that I encountered beyond *Spheres*. The men (predominantly) who think it's okay to talk about enthusiastically weaponising designers—that don't write down the female references in points of summary when co-teaching—that aim to shortcut discursive reflection to multiple-choice or think that collective student action happens in an immobile, patriarchal studio setting that operates panoptically, or like to explain to me about how my design world is organised around a sweet spot without understanding the implied "casual sexism" of these statements; would benefit from pondering indexicality (Bates 2012–). I get it, and it often gets to me deeply because it adds to the spherology that suggests we all occupy the same phenomenological inwardness, and the same sensory perception of being-in. What follows is not a list of correctives aiming to complete the *Spheres* index. On the contrary, the objective is to elucidate—as far as possible—what these gaps might mean to those that recommend readings for practitioners.

To "foam up" the index is to surface the bubbles that are submerged. To be clear, my "foamed-up" metaphor is not rabid (akin to a dog foaming with rabies), domestic (as an act of washing something clean, of rinsing the index as an act of domestication) or sexual (masturbatory nor orgasmic). Foamed-up just means scooping beyond the top layer of bubbles to view the indexical foam, i.e. what's indexically isolated. Narrative ideas in *Spheres* are often at the cost of a deeper indexical practice.

Who Cares About Indexicality?

While Sloterdijk was working on *Spheres*, feminist scholars like Isabelle Stengers were making central points on indexing (2005b). To appreciate Stengers' approach and contribution to philosophy for practitioners, it's not the theory that's so much the issue in academic writing but the authority or hubris that goes with it (ibid., 994). Stengers' earlier work on power, theory and authorship at that time investigated the normative practices of structuring science writing, predominantly bibliographic practices informed by the philosopher Denis Diderot's work on the encyclopedia (Stengers 1997, 145–60).

Other professions like archivists, marketers, librarians and publishers know the importance of indexicality. Take Michael Dieter, a new media scholar in Amsterdam, who gives some guidance to the constellation of disciplinary actants that may care. They are "researchers, theorists, hackers, and artists" (Dieter 2014, 216–7). However, my intention is to create a dialogue around an index that co-engages researchers, theorists, designers and artists (all of which could be hackers).

Sharon Webb's digital preservation work in Ireland calls for more digital research to "exploit" digital infrastructures already in existence (2015, 1). There's a gap, however, in this call. These digital access and storage systems in and of themselves may basically construct particular systemic issues before machine-learning enters the equation of index as algorithm. Academics continue to use seemingly uncontroversial and standard indexical methods. Furthermore, creating new and extended critical methods means enquiring into the standards of indexicality for digital scholars more generally. Yet, how many authors take the time to look at how diverse their indexes might be? I offer a way of approaching this gap when advocating for more digital research, but it starts with careful non-digitised reading because that's the media available presently to investigate *Spheres*. Understanding indexicality is not a research concern of past feminisms; it doesn't get "sorted-out" through open access or by extended machine-learning capabilities. As machine-learning becomes more prevalent in the indexical domain of books, these issues I raise may become even more hidden and complex.

How Global Is *Spheres*?

Readers have some way to travel (referentially) in the trilogy through a range of co-isolated disciplines, and for each, Sloterdijk gives short shrift to female authorship (see Table 4.1). While transgender is only notable in the Godly sphere it's not as a point for starting a trans-, poly-, binary- or fluid gender-oriented discussion.[2] Sloterdijk supports a narrative of heterosexual liberation, and supports procreativity, but without advancing the arguments made by women in relation to sexual liberation and reproductive choice.

Spheres includes texts in multiple languages and worldly sites of interest. In that sense, the trilogy is global, stretching laterally from Germany to India, and equatorially, from Africa to China. In relation to human subjects, its indexicality narrows to a singular worldview and the vistas

Table 4.1 Four philosophical texts and their corresponding number of citations according to the percentage of female authors per book

Author	Publishing date	Number of citations as per the trilogy index	Percentage of female authors in a book (%)
Sloterdijk	1998	$1117r = 49f + 1068m$	4.4
Deleuze and Guattari[a]	[1988] 2007	$595r = 39f + 555m + 1f/m^b$	6.7
Battersby	1998	$116r = 46f + 70m$	39.7
Stengers	1997	$113r = 7f + 106m$	6.2

r = reference, f = female author and m = male author
[a]The index for *A Thousand Plateaus* was compiled by Hassan Melehy ([1988] 2007)
[b]I have also counted deities and Krishna twice, for both male and female embodiments

and visibility atrophy and shrink to those of Imperialism when regarding gender mobility. Of those female names that make it into Sloterdijk's author index be they irregularly indexed (e.g. Saint Teresa of Avila) rather than absent (e.g. Christine Battersby), they represent under 5% of the referenced persons in both image and text. Is this indicative of other philosophers that he cites, or conversely that cite *Spheres*? Do the secondary literatures of *Spheres* reference any more or diversely than Sloterdijk?[3] These are the questions to be explored in the following sections and are lengthy because of the magnitude of evidence uncovered.

More generally, Sloterdijk's author index could do with a wider representation of female authorship and could be more equivocal in the fore- and sur-naming, which is compounded by the fact that the referencing of binary-gender varies. For instance, in the Suhrkamp index, Teresa is indexed by place name (by geography) "Avila" as "Avila, Teresa von" whereas John of Damascus is indexed by his forename as "Johannes von Damaskus". While Sloterdijk (like all indexers) is afforded discretion for parenthetical identifications, a symmetrical treatment of these two authors would have been preferential because deferring female names to their place makes them hard to find when attempting to search comparatively.[4] It's also an example of putting women in their place.

Who and What Counts in Spherology?

Table 4.1 compares the *Spheres* index with a snapshot of indexes from contemporaneous philosophers. The *Spheres* index was coded and double-checked in the Suhrkamp version, then checked against the Semiotext version—a total of four times.[5] Sloterdijk is the least gender-diverse philosopher when compared to Deleuze and Guattari ([1988] 2007) or Stengers (1997).[6] While the main narrative arc of Stengers' work is feminist, the gender ratio of the index is almost the same of Sloterdijk, which may say more about who gets to stay enlivened within philosophy's club of STS. To be clear, I'm not arguing that academics should publish more to achieve gender equality just as gender becomes simultaneously both more binary and fluid.[7] However, Battersby's gender-diverse indexing praxis indicates an active feminist approach to female authorship (PW). The same can be said of Wakeford in Table 4.2. I'm not suggesting that feminist academics tend to cite gender diversity as a *mise en equivalence*, because that isn't always the case. I'm suggesting that really fabulous feminisms work in both essentialised and non-essentialised registers, i.e. arguments based on feminisms with indexes to back that up, and feminist arguments that are multiple and inclusive for those who find binary or fluid gender relations supportive. Tables 4.1 and 4.2 are a numerical gesture for subsequent "citers" (i.e. readers that go on to write and cite) of *Spheres* in order to ask the question: is their reading and writing diverse enough? Though statistically less significant in their word length to accurately compare with Table 4.1 (based on their bibliographies rather than their indexes), the selected secondary literature (in Table 4.2) that cites Sloterdijk (which is not an exhaustive list) shows a slightly higher diversity of gendered referencing. That is, except for the philosopher, Carl R. Raschke, probably because he explores the heroic aspects of Sloterdijk's philosophy of religion, rather than a crisis of masculinity (2013). It doesn't seem that digital bibliographies, indexes or reference tools (like Bookends or Zotero[TM]) necessarily helped any of these second-generation writers to rethink Sloterdijk's social and philosophical arrangement of literature. These tools at present are limited in regard to intersectional issues. When contrasting how particular disciplines reference and how philosophical ideas travel into other disciplines through citational networks of enquiry, the attention to gender remains important.

Table 4.2 Seven secondary texts related to Sloterdijk's concepts of *Spheres* and the previous related ideas contained in the *Critique of Cynical Reason*. Percentages of female authors are listed per book as related to the citations or bibliographies for the journal papers and book indexes, if available for the books listed

Author	Publishing date	Number of references	Percentage of female authors (%)
		By journal paper	By journal paper
Houdart	2008	40r = 5f + 35m	12.5
Raschke[a]	2013	8r = 8m	0
Tuinen	2010	14r = 1f + 13m	7.1
Tuinen	2009	16r = 2f + 14m	12.5
Wakeford	2011	13r = 4f + 9m	30.8
		By book index	By book index
Couture	2016	62r = 3f + 59m	4.8
Elden	2012	152r = 7f + 145m	4.6

r = reference, f = female author and m = male author
[a]The list of references are taken from the footnotes because there's no bibliography

When comparing the two female philosophers, Babette Babich and Marie-Eve Morin's referencing practices (SN), Babich alone refers to the work of three of the seven women named in the entire edited collection by Stuart Elden (Babich is the third reference). Morin cites no female authors in relation to Sloterdijk's work because her frame of reference repeats the notable philosophers contained in *Spheres*: Georg Wilhelm Friedrich Hegel, Martin Heidegger, Lao-Tzu and Jean-Paul Sartre. For Morin, the concept of "Dasein" in her chapter about *Spheres* and environmentalism is merely Sloterdijk's rethinking of Heidegger's concept of throwness—of how a child understands how to alter the exterior world by throwing a stone. Morin's text could have been a bridge to feminist forebearers and activated female thinkers crucial for Sloterdijk's notion of environmentalism and globalisation in *Spheres*. For Morin, the "cage" of being is environmental and the humanness is indescribably animal and anthropological—of bone (ibid., 81–4). Morin has yet to throw a stone and cite a female reference, let alone a feminist thinker from the East. South Korean novelist, Han Kang, and her book, *The Vegetarian*, would push fiction back into philosophy, something of which Sloterdijk advocates as a creative method ([2007] 2015). This fictional story can show how women can twist and physically reshape themselves to their so-called "natural environment" within a dominant patriarchal sphere, as a matter of mental health.

In the next section, I investigate if there's a more active and sped-up process to tackle "citation issues" that doesn't require an academic to absent entire categories of being-human (like eradicating white men as Ahmed does) because if that were the case I'd be inclined to unreference Morin, but that's unethical.

IMPROVING GENDER DEFICIT BY A 1990S TECHNOLOGICAL PREDICTOR

Let's imagine that an academic generation comes along every ten years based on hiring and firing practices related to research assessment exercises and tenure-track systems (including fixed-term and temporary contracts), rather than on birth and death rates. It would take centuries for citation practices to become more gender-diverse let alone to catch up with more radical diversity practices.[8] To play with numeric gestures and to anchor individual counts to one of those measuring tethers; if digital publishing rates were predictable following the same exponential increases of Moore's Law, then the rates of female referencing would accelerate accordingly.

In Moore's Law, the number of transistors in a dense integrated circuit was predicted to double every two years which seemed correct for almost a decade. If instead of stating transistors and circuits this was to be replaced with female authors in a digital index, those predictions would have meant that indexes would have reached a gender equilibrium in under a decade: 1998 (4.4%); 2000 (8.8%); 2002 (17.6%); 2004 (35.2%) and 2006 (70.4%). Also, his exponential rate of increase would have effectively countered gender imbalances within a decade and within the same timescale that Moore's Law remained correct as a predictive measure. In around six years, the referencing rates would have matched that of Battersby or Wakeford to 39.6 and 30.8% respectively (see Table 4.2). It is *only* feminist writers making an intervention into citation practices that could get near to the digital predictive measures of equivalising at a doubling rate. There's of course, no regulation for gender parity in indexing rates because it would seem absurd to measure human values (as binary-gender) by technological prediction.

On a larger scale, what does exist in a study by Elsevier (2017) are the quasi-global counts of women researchers in 17 countries, not including China, India or other Pan-Pacific countries. The Gender Research Benchmark Report shows that from 1997 to 2017, 49% of all researchers are women, which is an 8% increase since their last report (1996–2000).[9] In this study, as summarised in a review by Alice Meadows (Director of Community Engagement at ORCID),[10] the gender gaps in publishing display an unfolding landscape of differences regarding leadership, collaboration, interdisciplinarity and mobility. *Spheres* is strikingly similar in geographical range as the Elsevier report, yet does not convey gender diversity (Table 4.3). The report's overview states, "Although women tend to publish fewer research articles than men, their articles are cited or downloaded at similar rates", and I infer that *Spheres* rather than the discipline of philosophy is the exception. Sophisticated digital tools are not essential to show feminist researchers the discrimination against their bodies. Digital researchers ought not to contribute to new methods that sustain these un-diverse practices. It's about noticing what Battersby and Wakeford do implicitly and replicating it across disciplines.

To do so means adopting a strategy from Stengers' etho-ecological essay, that is, to slow down and hesitate (2005a). To be clear, I'm not arguing that we should publish more to do this; on the contrary, to slow

Table 4.3 The table shows the countries of birth and/or residence for all of the female authors in Spheres. The countries refer to the places where the indexed female actants were born and live(d), that is, if these are different locations, or indeed known. When a reference appears in volumes, the count is made more than once. The only female to appear in more than one volume is Diana Princess of Wales, not even Luce Irigaray or Julia Kristeva count more than once

Member States of the European Union (in 2017) by place of birth and residence	*Total*
1. Austria	7
2. Belgium	2
3. Bulgaria	1
4. France	8
5. Czech Republic	3
6. Germany	12
7. Italy	3
8. Spain	5
9. Sweden	1
Non-European Union (in 2017)—by place of birth and residence	
1. Barbados	2
2. Canada	2
3. Japan	1
4. North Africa	4
5. Prussia	2
6. UK—withdrawal process from the European Union started in 2016	7
7. US	22
8. Yugoslavia	1
9. Unknown	16
Total	99

down enough to index diversely (aiming above the threshold of 30% like the texts by Battersby and Wakeford) thereby speeding up diversity rates rather than publication rates. There's a paradox for those entangled in the publication practices of academia. The issue here is that those citing diversely, I'd hedge, do not publish as much as those that couldn't care less. As most early-career researchers strive to publish more, I wonder at what cost? It's easier to know about terrible citation practices than it is to resolve these complicated matters inter-generationally; but in the next section, I'll show why it's important to try to do something about it rather than just notice it and leave it at that.

The Forgotten and Absented Women
of Incarceration and Death

By interlacing Sloterdijk's 12 endnotes, 18 captions to images and 24 references that finally make it into the main text (of 54 indexed citations) I'll begin, volume-by-volume, to knit together the isolated references and intertwine some of the theories of those female authors that are obscured or missed in *Spheres*. To peruse the deathly and depressive content of *Spheres* is to assume that women have little or nothing to say about these states of being—that they do not act, contribute or write on depressive objects and their circumstances.

> However, this heroic posture is not without a certain number of blind spots or political blunders, which perhaps reveal the philosopher's desire to become the prince's advisor, to trade in the nightingale's feathers for those of the hawk.
>
> —Couture (2016, 75)

With a quote from Babich, I will "return to the sexism that never really left" to counter Jean-Pierre Couture's above point about trading death for war, and depression for predation (SN, 33). Couture seeks forgiveness from his readers for the gaps and errors in Sloterdijk's work. I'm hawking a media-theory star (Sloterdijk) by detailing the issues which a researcher can uncover: the blunders by doing minimal research—little more than just an encyclopedic check. The count of 50 women in *Spheres* needs adjusting (to delete Chessman and add Haraway and Margulis) in the index.

In relation to Bubbles I: Sloterdijk's argument relies on patient narratives whose stories excite volume I, with little regard to indexing their contributions because Sloterdijk cares little for expert-patient accounts. The identical twins Jennifer and June Gibbons are two of the many patients whose stories explicate on the trope of bubbles. The two autistic twins lived as children in Wales, having moved there previously from Barbados. They shared a private language and were both prolific journal and fiction writers in their own right. On accounts of arson and other crimes, they became patient-inmates at Broadmoor in the south-east of England. They appear in *Spheres*, indexed as a singular citation, appearing in the English Semiotext edition as "Gibbons, Jennifer und June", indicating that the index was not amended when translated from German to English; therefore, the errors aren't those of the translator (Wieland Hoban), but the unnamed indexer. The reference does not follow the ISO 999 standard indexing of two or more persons with the same name.

> Two of more persons of the same name should be distinguished by the addition of qualifying information, such as dates, occupation, or title.
>
> EXAMPLES:
> Butler, Samuel (1612—1680)
> Butler, Samuel (1835—1902)
>
> —ISO (1996, 15)

It would seem that Sloterdijk's ideas may either be from *Silent Twins* (1986), the book, or the subsequent television film (1986) written by the journalist, Marjorie Wallace, with the film directed by Jon Amiel, but there's no reference for this content. It could be that Sloterdijk read the *New York Times* review of Wallace's book in which Sacks recounts how Wallace contextualises the creative writing and illustration of the twins, which includes diaries, poems, novels, stories and illustrated strips (1986).[11] At that time, both twins were alive and had succeeded in publishing independent works like June's *Pepsi-Cola Addict* (1982) and Jennifer's *The Pugilist* (out of print but cited by Wallace in 1986). Sloterdijk takes the twins' life history only for their unusual autistic behaviours which strengthens Sloterdijk's argument about "doubling" (the sharing of an intimate communicative bubble or sphere), an emphasis Wallace makes herself, which is made clear in her later article from 2003.

The twins published independently, but Sloterdijk refers to June and Jennifer as a case study of one shared, linguistic sphere, strengthening the sisters' inseparability; yet their story is about being torn apart. They were separated several times in their institutionalised lives with dire consequences. The twins talked about how only one of them could survive the other (Sacks 1986). At the time of the first publication of *Bubbles (I)*, Jennifer had already died. Concerning the other younger sister Rose Gibbons, and also June Gibbons, Sloterdijk could have indexed the twins as two separate individuals as an indexical act of ethical compassion, even if it runs counter to his argument about doubling, biuneness and coupling. What is strikingly different in the writing of Sloterdijk than that of Sacks and Wallace is compassion. It's remiss in *Spheres*. Sacks explains how Wallace's book is thoroughly researched, investigative journalism.

Considering it populist news, it is no excuse for Sloterdijk's [-in]ability to attend to indexicality. Authors are meant to follow citational standards so as not to passively extend their own beliefs—in this case, of doubling.

Sloterdijk does index fairly most theological citations, but that's odd in itself. For instance, Sloterdijk distinguishes the saints of Carthage, Felicity and Perpetua, but why? Does he do this to note their differences—to recognise slave from free women or ethnic difference, or is Sloterdijk indicating a difference in their motherly sainthood? Felicity is the patron saint of widows and mothers of deceased sons whereas Perpetua is the patron saint of mothers, notably expectant mothers, ranchers and butchers (and of both Carthage and Catalonia). Sloterdijk's interest with deceased mothers (I and II) is an underpinning theme that resonates in the indexicality of *Spheres*.

In relation to Globes II: Queen Elisabeth II of Great Britain, appears twice in the index, yet, the reference in the text is related to the Queen of Spain, Elisabeth Farnese (II, 59). This may be a careless error, but not indexing by their titles which he does for Saints shows how Sloterdijk upholds the traditions of titles in theology but not for female sovereignty. This error shows Sloterdijk's underlying bias towards indexing. The trilogy names: Princess Diana of Wales (II, 732, 601, 784), Marie Antoinette (ibid., 574) and Marie-Anne Charlotte de Corday d'Armont (ibid., 571).

> Persons normally identified by a title of honour of nobility should be indexed under that title, expanded if necessary by their family name.
>   EXAMPLE Boudicca, Queen of the Iceni.
> —ISO 999: (1996, 15)

Sloterdijk could have argued against regal patriarchy to debunk sovereign power, but rather he simultaneously underplays and de-contextualises the biographies of these women to emphasise their relation to death, murder and assassination. All of this leads to emphasising a female depressive sphere.

Regarding locations of the women indexed in *Globes (II)*, these women were born or resided in at least 17 countries worldwide (Table 4.4). While this may be an obvious point to make, the literature cited in these countries falls along the lines of the trade routes of empires and paths of mobility granted to missionaries. It is through the lack of women's conceptual contributions to be found in *Globes (II)* where

Table 4.4 The total count of references as they appear in multiple across the trilogy, i.e. in the content, image and endnotes would count thrice. This is why the number of women identified in *Spheres* (54) does not tally with the total number of female authors cited (49) and the overall reference count of 65 (i.e. $11 + 54 = 65$)

Volume I			Volume II			Volume III		
Content	Image	Endnote	Content	Image	Endnote	Content	Image	Endnote
10	6	2	6	0	5	8	12	5
	Subtotal 18			Subtotal 11			Subtotal 25	
							Total 54	

heteropatriarchy gets writ large. Furthermore, the dramatic narratives are based on borderline personalities.

In relation to Foams III: Spheres relies on individual stories of state injustices with prisoners. A grave error in the index (and repeated in the prose) is the naming of Cheryl Chessman as an example of a Californian state execution by cyanide in 1960 (ibid., 110 and 892). The event strengthens Sloterdijk's argument around design and the trope of air. When reading the numerous contemporaneous articles on Chessman's execution and prosecution, Chessman is, in fact, Caryl Whittier Chessman. It's a tragedy to note in this additional literature that Chessman was indicted by the California state in 1948 on 18 counts of robbery, kidnapping, rape and murder. It is also a tragedy to learn that Chessman's state execution took nine minutes and during that time a phone call was received to stay the execution. Out of respect to both the unnamed and un-gendered victims, and Chessman himself (a victim of state execution), this mistaken gendered identity needs renaming when citing the case of Caryl Whittier Chessman.

In some respects, the Chessman example is a small typographic error, but the use of the patients and prisoners of colour, including minors, is a deeper issue. It is through compassionate prose and citation practice that gives dignity to those detailed into any account of the depressive, tragic or macabre. The register needn't be inflammatory because words can heal. The method underpinned by Kynicism (i.e. declaring the falsehoods made by cynicism) is, I'd argue, oddly usurped by so many falsehoods appearing in *Spheres*. Sloterdijk leaves intact an "atmosphere" of colonial confidence to his global topicality, which, I'd suggest, underwrites a heteropatriarchal voice that underpins the notion of "world-leading" scholarship. While I've attended to the missed, obscured and absented, I'll next consider what Sloterdijk makes of the "strong" female references to be found in *Spheres*.

THE "STRONG" PSYCHOANALYTIC REFERENCES TO WOMEN

When the references are extrapolated in the ways previously mentioned, it's really hard to hold onto the pages where very few women are referenced. There is only one "strong" female reference and that is for Julia Kristeva. While physicist and chemist Marie Curie, psychoanalyst Luce Irigaray and philosopher Hannah Arendt do feature in the main text, they are still fleeting references. Kristeva, a Bulgarian-French philosopher, literary critic, psychoanalyst, feminist and novelist will be familiar to artists and art historian scholars for her work that spans three decades. One of the longest quotations of the trilogy is given over to Kristeva's own words occupying three paragraphs across three pages (I, 528–31). References to Kristeva's art history are absent and her words confuse rather than enlighten the beginnings of *Spheres*. It's a missed opportunity for Sloterdijk to build a joint readership in performance and installation art with literature (see Chapter 5 for more).

I think the desiring subject for Sloterdijk is women. He tends to talk about women as containers, as deceased mothers, as widows, as the interned, as maddened, murdered and raped but in order (indexically) to transcend them. It is womanliness in all the deadliest of convictions. In this sense, Sloterdijk writes from an addictive prerogative going in search of new content rather than attending to issues with the indexicality of *Spheres*. While Sloterdijk may have been deeply affected in ways the reader is not privy to knowing, I will say this: Sloterdijk has struggled with gender in *Spheres* BUT he could have understood his impetus to "write out" and could have referred to Hélène Cixous' text, "The Laugh of the Medusa" (1976), whose essay helps to challenge gender inequality through the action of writing. Had he slowed down and drawn more from both genders in relation to art, design, philosophy, religion, literature and history, and had he read Cixous as well as Irigaray and Kristeva, the depressive in *Spheres* may have been quite a different read. He cites too few feminist writers to understand feminist methods of care and he sidesteps the Kleinian psychoanalytic concepts of care related to "reparation", "splitting" and "natalism" (Klein 1929; Frampton 2004), but perhaps this would have deflated the trilogy's unintentional main purpose—to creatively explore the depressive sphere as a form of female sexual promiscuity, or female sexual liberation. The distinction is unclear. What is clear is that out of 49 female references appearing across all volumes, only 12 appear in the endnotes. That means that 24% do not appear in the copy of the text, either in the prose or the captions to images. This could suggest that these women are not pivotal to Sloterdijk's main argument and are peripheral, but is this the case? On the contrary, there are (as I've been pointing out) quite a few writers that are more pivotal than the term "endnote" suggests (like Haraway). The paradox of Sloterdijk's endnotes is that often these feminist authors don't end an idea, but instead they catalyse an alternative spherology to Sloterdijk's *Spheres*.

Sloterdijk's irritated attempt to glean "factual" information about women as simply distinguishable body parts is derived from a few writers like Barbara Walker—a feminist who has written broadly on topics from religion through to knitting.[12] At some level, Sloterdijk is extending the psychoanalytical modes of writing about female body parts (i.e. vagina and vulva) to understand female phenomenology, yet he continues to refer to female body parts within a concept in masculinised containment. It's exacerbated by Sloterdijk's attempts to explore the female body as encyclopedic through definitions of the "feminine" as a word search. Sloterdijk seems "irritated" by how an encyclopedic search cannot supplement or round off his ideas for what is deemed female, via deterministic definitions.

> One can see how it is possible to pass by this entirely, not least under the pretext of presenting a cultural history of the feminine, by reading Barbara G. Walker's more irritating than useful Women's Encyclopedia of Myths and Secrets (New York: Harper Collins 1983) in the hope of learning something about keywords such as birth, fetus, initiation, placenta, return, search, separation, vulva, etc.
>
> —Sloterdijk (I, 645 n7)

Sloterdijk appreciates an encyclopedic read, but a female reader may not appreciate *all* of his encyclopedic endnotes. For instance, rather than her single-authored books and poetry (i.e. 1994), Sloterdijk finds the feminist writer Helga Häsing advantageous for her work on "separation" and "loss" from the *Feminist Encyclopedia of German Literature* (1997 in III, 886 n138), and also for her joint work with psychoanalyst, Ludwig Janus entitled *Ungewollte Kinder* (1994). To follow a trend, neither references are in the main text, but rather prop up a pre-natal-ist argument stemming from Thomas Macho's account of psychoacous-tics from the womb (prevalent in I). So far, I've highlighted the ways Sloterdijk searched encyclopedically as a mode of knowing "women" and how women are preferentially cited as co-authors, rather than single-au-thors. There's a continual pattern that emerges: Sloterdijk prefers to cite white women as co-authors rather than single-authors. The indexicality of *Spheres* is therefore categorically co-distorted and doubled.

CO-DISTORTIONS AND THE MATILDA EFFECT

What's going on in *Spheres* are "citation distortions" meaning that Sloterdijk has claimed for himself and other male authors more authority than he should. Historian of science, Margaret W. Rossiter, calls this the "Matilda Effect" (1993) which specifically looks at the denial of contri-butions made by women and that are attributed to male authors. Häsing, among others, are further evidence of this effect (II, 1018 n133).

The further distortions in *Globes (II)* are complex and again grounded in researching the endnotes. They refer to the politics of globalisation and the trope of globes. Karla Poewe, Martha Craven Nussbaum and Christa Müller (and their co-authors) are the pivotal globalisation ref-erences in *Spheres*. They add to the politics of globalisation from social democracy (and Social Democratic Party, SDP), the economy and a "free market religion".

Poewe has published numerous works in anthropology including fem-inist writing on the *Universal Male Dominance: An Ethological Illusion* (1980). Poewe left East Prussia as a refugee during the second world war and moved temporarily to Poland before subsequently moving to Canada and then academic of South Africa and the Zulu African Independent Church. Poewe married Irving Hexham and the two of them co-authored *Understanding Cults and New Religions* (1986) and *New Religions as Global Cultures* (1997). Again, like Häsing, it is Poewe's co-authored work with her husband that is referenced by Sloterdijk, rather than her single-authored writing which was first published several years before *Globes (II)*. It is from this packed endnote that Poewe's writing relates to Sloterdijk's point about the harmonising of the developments of the free market of religion.

Karow taught Religious Studies at the Free University of Berlin and studied the Nuremberg trials including the cult of Bhagwan Shree Rajneesh (Stackelberg 1999).[13] Both Poewe and Karow are publishing work on the cult that Sloterdijk took part in; that is, if he did more than just observe this cult as an anthropologist-philosopher (III, 873 n138). It's not odd at all for Sloterdijk to be reading about the very cult he was a member of yet the potential to conceal or even obscure such a reading is un-biographical. It's highly likely that Sloterdijk has read (if not been interviewed) by Poewe and Karow, but I can only speculate because this is to pursue the undoing and reversing of consent. To my knowledge, Sloterdijk is writing tacitly (guarding knowledge) about these experiences and it adds to an argument that *Spheres* is an autobiography that aims to fragment as well as un-biograph himself to the female body. He is thinking and writing something through, which is evident in citing Karow and Poewe, even if in the shades of an endnote that cannot be fully deciphered by a curious reader.

Müller, until 2011, was a family policy spokesperson for the Left in Saarland and an influential politician for the SDP (Social Democratic Party of Germany) and Die Linke in Germany. Christa Müller is again only cited for her joint work with her husband Oskar Lafontaine on globalisation (1998). I could argue that Sloterdijk utilises the political speeches and writings of Müller more often than what gets cited in *Spheres* (ibid., 1018 n133). I'd hoped that Sloterdijk would discuss some of his political media appearances and how he's influenced German politics via the media since 2002. This could have led to some interesting autobiographical sections in relation to globalisation and politics but Sloterdijk chooses instead to give un-biographies.

The broadest, most global ideas do extend female authorship but through marital status. This is a concerning way to strengthen ideas about foam worlds but was it intentionally more subversive than subconscious? If he'd stated a point about how no one writes alone and that writers think through more than they write, this could have been generative; but instead, *Spheres* strengthens his ideas that women are mostly second-rate authors in *Spheres* (TTM, 3).

Nussbaum is cited by Sloterdijk for her writing on "Aristotelian Social Democracy" in the co-edited book entitled, *Liberalism and the Good* (Nussbaum 1990), in II, 1018 n133). However, as a prolific American philosopher and an Ernst Freund Distinguished Service Professor of Law and Ethics at the University of Chicago, Nussbaum also held a joint appointment in Law and Philosophy and has so much more to offer a reader of *Spheres*. During her studies at Harvard University, Nussbaum talks about her issues with excessive discrimination, sexual harassment and for problems getting childcare for her daughter (1997, 7). Nussbaum's notion of social democracy derived from classical studies aims to move academic debates and ideas of democracy by instilling in prospective students the concept of "intelligent citizenship" (ibid., 11). While both Nussbaum and Sloterdijk focus on religion, there are other points to draw on from her literature, especially regarding Nussbaum's concepts of "democracy" (ibid., 6). In the same text as previously mentioned, Nussbaum discusses her position within academia recounting her time as a fellow and how she hoped to further "cultivate" rather than "produce" citizens in students, which means not just voting but being able to participate in those debates. Sloterdijk skews Nussbaum's feminist, pedagogic and theological points. I'd hoped that Sloterdijk would discuss some of the hazing and harassment he must have potentially witnessed both in academia but also during his time in the cult, in Pune, India. Perhaps it could have led to alternative sections on globalisation and authorship, the politics of academia and maybe a complete rewrite of the phenomenological treatment of women as a container and "female" as a concept defined by distinct body parts (to be understood through encyclopedic entries). Sloterdijk made choices between prose and endnotes, repeatedly choosing to cut out the women or tie them to bonds of co-authorship and marital status, particularly when in the global sphere. This argument leads me to consider "discipline distortion" as a sort of foamed club where there's little dialogue between feminists' ideas. Kindly check the usage of the parenthesis in the sentence 'Nussbaum is cited by ...'. Amended.

From Male Single-Authorship to Heteropatriarchy Interdisciplinarity

Artists and designers may not care about the count or to reference quite so instrumentally as I have done, but I cannot ignore the counting that's so prevalent in the autobiographical research of the "quantified selfers" and the "big data visualisers" of the algorithmically quantum-coded twenty-tens.[14] In the same essay that Stengers makes a point about disciplinarity (mentioned in the introduction), she also makes a social point at an institutional scale. Stengers argues that disciplinary models of work in the sciences are akin to "hunting in packs" (ibid., 130). This way of knowing is a predatory form of collaborative survival. In philosophy, the Frankfurt School is one such "pack" of which Sloterdijk was affiliated too, but in recent years he theoretically stands against it, aligning with world-leading design schools at Boston and Delft.

DISCIPLINARY DISTORTION IN *SPHERES*

At the time of writing this book, both the Suhrkamp and Semiotext versions of *Spheres* were not available digitally so researching disciplinarity meant relying on handwritten notes and coding. In the next section, my argument is limited to the most problematic of the volumes: I–II. The indexes by discipline were created by speed reading volumes by hand, by four separate searches, whereby my initial searches were put to one side and started over and then compared and checked for gaps. Before going on to use the physical shape of the metaphors of bubbles and foam in diagrammatic form, it's worth noting that in *Spheres*, bubbles is assembled from female geometry, yet masculine when scaled-up to globes but less so when pluralised into foam. The metaphors are not as gender-neutral as they first appear. While in some secondary literature these metaphors and tropes become more inclusive, all three tropes of *Spheres* have yet to trans-end gender inequality in metaphorical form—a potential premise for a spherology of feminisms.

The diagrams shown in Figs. 4.1, 4.2, 4.3 and 4.4 are gender-neutral visualisations because the tropes are not tethered to the gender-biased

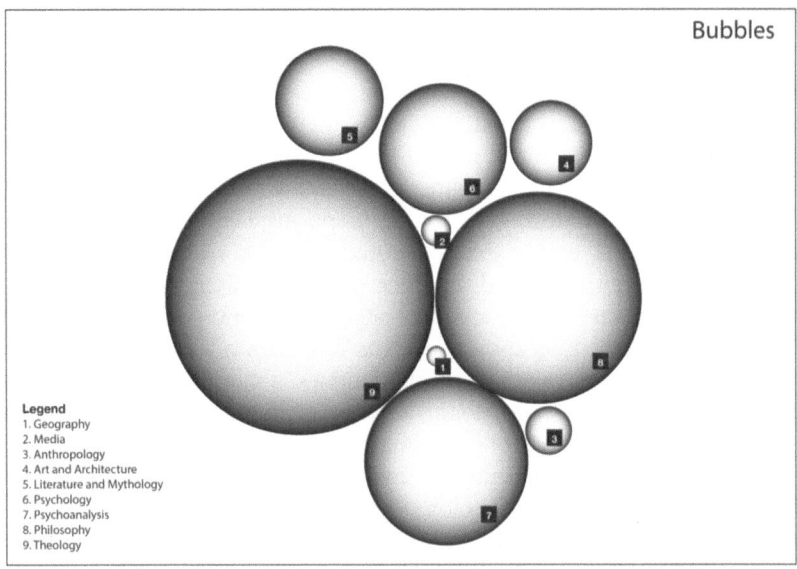

Fig. 4.1   A visualisation of the proportion of *Bubbles (I)* (*Source* Windle)

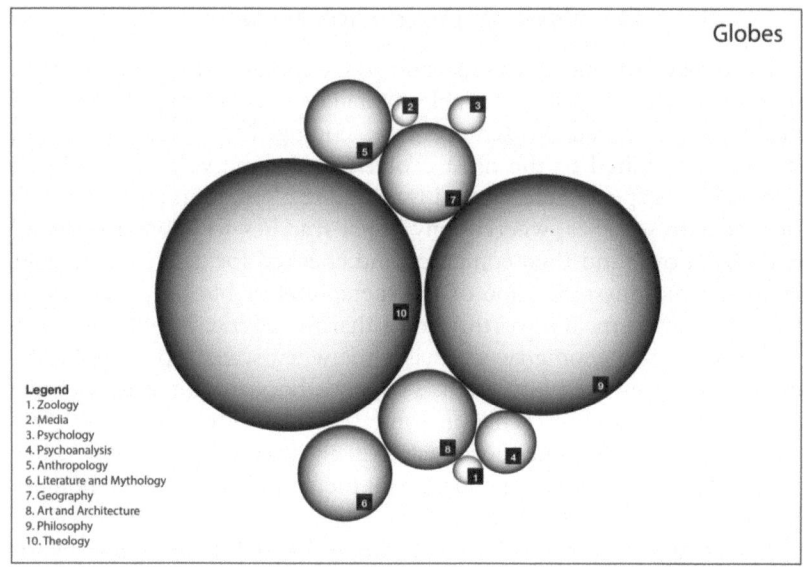

Fig. 4.2   The bubbles are proportional to the range of disciplines represented in Sloterdijk's *Globes (II)* (*Source* Windle)

narrative or indexicality of *Spheres*. Designers can break with the heteropatriarchy especially when *Spheres* is oriented to design, which is to come, but there are a couple more I need to make first.

Sloterdijk cares about disciplinarity, referring to disciplines specifically in the text. Each instance whereby he mentions a discipline was indexed and the length of each extract I noted to its corresponding page(s). These pages were then added together to create a discipline's overarching length. Images were also counted as page-length and added to the totals by area. Disciplinarity is shown as areas in circles organised by size (see Figs. 4.1–4.2). These take Sloterdijk's simplest spherical form of bubbles. The final Fig. 4.3 draws on foam, the most complex geometry to illustrate the range and spread of disciplinarity, covered in volumes I–II.

A statistical researcher may be inclined to group the disciplines in *Spheres* by combining psychology with psychoanalysis, or geography with zoology, but Sloterdijk chooses to use these terms independently and so they're left as read. Figure 4.2 shows two clusters of disciplines that are similar in proportion to one another, giving the appearance of

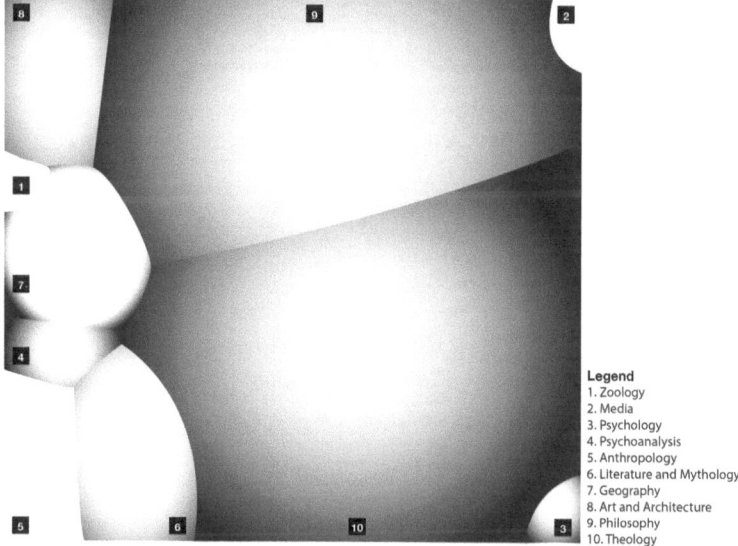

Fig. 4.3 The shape of each bubble is proportional to *Globes (II)*—the macro-spherology. The shading of each bubble (i.e. the darker the shading, the more prevalent a discipline) is proportional to *Bubbles (I)*—the microspherology. The Voronoi were initially made using Opensource, Voronoi 3D modelling software by inputting a Microsoft Excel™ .csv file. These initial visuals were reconstructed using Adobe InDesign™ (*Source* Windle)

symmetry between disciplines. Bubbles of equal size are labelled with the same number; for instance, media and zoology are given the same proportion of space on the written page. Bubbles at the top of Fig. 4.2 are labelled 2, 3, 5 and 7 (media, psychology, anthropology and geography) and get bigger in relation to the bubbles at the bottom marked 1, 4, 6 and 8 (zoology, psychoanalysis, literature and mythology, as well as art and architecture). While geography is prevalent in *Spheres* I–II, the discipline is more common in the discussion of globes, hence why zoology turns up only in the second volume.

It would be naive to infer that the disciplines that Sloterdijk has chosen to acknowledge are as heteropatriarchal as his index by name would infer. Though, if Sloterdijk is citing only male leaders within each

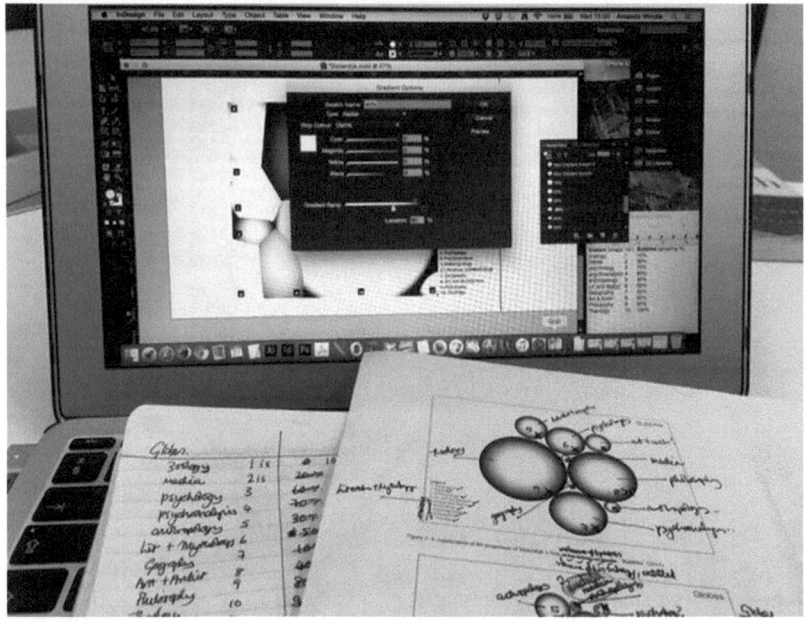

Fig. 4.4   The computer screen shows the adjustment of shading to the bubble forms (editing Fig. 4.3) by increasing the white in the colour percentages of the gradients window of Adobe Photoshop™ CC. The gradients refer to the hand-written notes made on earlier versions of Figs. 4.2 and 4.3 (bottom right). These were then adjusted first by hand (bottom left) using TextEdit™ (bottom right of computer screen) in conjunction with Photoshop™ (*Source* Windle)

discipline, the following trend in the Elsevier report would be correct (2017, 41). The trend suggests that academic authorial leadership has been developed by distorting female authorship through shared practices of citation, and that gets augmented at the scale of disciplinarity.

How Is the Trilogy Illustrated?

To illustrate the theory of foam there is one image that stands out in *Spheres*. It's a reproduction of Rene Descartes' 1644 Voronoi diagram of star systems (II, 123). The Voronoi depicts the relations of constellations of stars in the sky. Before explaining why a Voronoi is the data visualis-ation of choice to depict foam, I'll first introduce the variety and oddity of imagery to be found across *Spheres*. The trilogy is beautifully and prolifically illustrated with European paintings, sculptures and engravings predominantly taken from Sloterdijk's collection of photographic reproductions. To a lesser extent, there are scientific models, diagrams and images of earth taken from space. There are two-dimensional charts, atlases, maps and three-dimensional globes, as well as machinery for heart surgery and anaesthesia, fetal ultrasounds in the womb and drawings of medieval gynaecology. There are anthropological artefacts like Neolithic pots, coins and pottery from antiquity.

Unlike the other philosophers that Sloterdijk cites, he never illustrates his ideas himself. Neither his seven-layered microspherology (I, 540) nor the "dynamic foam axes" of his nine-dimensional space are accompanied with visualisations (III, 464). Other philosophers like Gilles Deleuze and Felix Guattari do illustrate their ideas, like their "six lines that make a complex machine" ([1988] 2007). However, Sloterdijk opts for sketches by others like scientific photography of stars from the Sculptor constellation which are taken with the Hubble telescope, and blackboard drawings by the conceptual artist, Rudolf Steiner (I, 22 and 578).

Since *Spheres* was first published, contemporary diagrammatic practices in visualising data tend to require a mix of hand-rendered and digital methods. For instance, designer Stephanie Posavec's maps, in part, detail categories like "women, sex and relationships" as a key thematic of Jack Kerouac's novel, *On the Road* (1957). The maps are coded by hand, then digitally rendered, similar to the way I "marked-up" *Spheres* (2004). Alternatively, Suzanne Treister's "Alchemy Project" hand draws content from news which questions the prevalence of graphics in current pictographic news stories (2007–2008). Or in 2015, the Nesta Quantitative Research Fellow, Catherine Sleeman's animated online data visualisation work, accompanies a narrative about "Antimicrobial resistance across Europe", which is co-authored with the Digital Content Editor, Nina Cromeyer Dieke.

For other quantitative data—techniques more aligned to networks and Actor-Network Theory (ANT) rather than a theory of *Spheres* see—Noortje Marres (2005), Noortje Marres and Richard Rogers (2005) as well as Geneviéve Teil and Bruno Latour (1995) among others like Alberto Cambrosio et al. (2004). There is a prolific use of relational network mapping in STS, but less so regarding the mass of a spatial area, which is more easily depicted with *Spheres*. STS scholars seem to have distanced themselves from the count by area because it's an oversimplification of the social as a container, which runs counter to the "nets and weaving" of ANT, STS and feminist technoscience. Sloterdijk doesn't lead by sketching, diagramming or drawing because he chooses to remain in the written sphere of arguments and ideas, rather than inspiring a model of practice-led philosophy. In the next chapter, I look at the work that comes after *Spheres*—that extends this discussion of design practice—but for now, the argument of choosing a Voronoi over a Venn needs concluding.

CHOOSING A VORONOI OVER A VENN: WHICH SPHERICAL DIAGRAM MATTERS?

Venns are so prevalent in design presentations and reports that they are too numerous to count. Voronois (like Fig. 4.3) have a characteristic foam appearance. However, Sloterdijk's visualisation appears in *Globes (II)* rather than *Foam (III)*. A voronoi's foaminess is helpful in showing both spatial mass and relational connections (more prevalently illustrated in network diagrams connected to ANT). The Voronoi is often used as a layer over a landscape or vista in astronomy, ecology, geography, meteorology and mining. However, he doesn't refer to it precisely as a Voronoi diagram. There is no debate about diagrams for designers in *Spheres* but it warrants more than one which I'll now provide.

Figure 4.3 depicts the proportionality of disciplines through structural tension. Each bubble-cell's structure (in Fig. 4.3) is held stable by each bubble's spatial relation to another. The positioning of each cell in relation to a corner of the Voronoi is arbitrary. Had I mapped each disciplines' placement within each section of the trilogy (page numbers not just page length) this Voronoi would then have mapped the travel through disciplines that Sloterdijk takes. The Voronoi would not just show disciplinarity by area but the relations between disciplines, notable by the cellular walls. However, this method would not be able to distinguish in *Spheres* between a co-related or co-isolated disciplinarity.

What of the Venn? When two circles or more intersect (to make a Venn diagram), the two overlapping arcs make an ellipsoid or petal shapes. It's often referred to in presentations (by designers) as the "sweet spot". This visually represents the main point of any presentation. The logic of production is sexualised and revealed like the discovery of a female orgasm. My main point is that the Venn diagram becomes a sensationalised-material form and the point becomes sexualised. I've really had enough of designing them because I've witnessed too many "everyday instances of casual sexism" to use the title of Laura Bates' Twitter™ feed (from 2016). Casual sexism occurs when Venn diagrams are "mansplained" or sexsplained: when a person explains something to a woman that they already know, in a way that indicates to women that "this is not their world" (Solnit 2015, 4). In other words, and to parody with irony and binary-genders, Sloterdijk's being-in formula (TTM, 7) from Chapter 1: *Being-in-a-diagram would mean a female designer (1) being [explained how] to be together with male designers in general (2) with casual sexist remarks (3) whilst in a Venn diagram (4). This formula describes the minimum complexity a female designer needs to arrive at an appropriate concept of design worlds. Female designers are involved in this consideration since for them being-in-a-Venn means dwelling in a diagram [to add] and as a Venn, this is a female place from which this mansplaining comes from, if not to come.*[15]

Conclusion

Just because Sloterdijk is at ease with many languages—moving across cultures in the narrative prose—does not mean that his indexicality is representationally global or world-leading. My visualisations may seem to show that Sloterdijk's two volumes are diverse by discipline but when looking closer in the preceding sections of this chapter, it is not that diverse on an authorial basis because there are far too many forms of distortion.

To conclude, the ideas and subject matter of *Spheres* are co-constituted from 1068 men (not including Chessman) and 49 women (not including Margulis and Haraway). It is these ideas that constitute Sloterdijk's theory of *Spheres*. Yet, scattered and co-isolated in between the majority of ideas are crucial feminist texts that constitute a spherology of feminisms. If you are aiming, as I've done, to make a point through design practice about indexicality, then be warned that to do so will furthermore delay and slow down an academic's ability to publish at speed. However, the developmental point (I'd like to think), leads to a more vivid, methodical and deeper understanding of an alternative spherology. And once again to parody (as a rather lengthy stanza for this chapter) Sloterdijk's abstract definition of being-in (ibid., 7): *The Spheres index means a design reader (1) is together with a few female references (2) and with a predominantly all-male index (3) situated at the back of the third volume, physically apart from the first two volumes covering a*

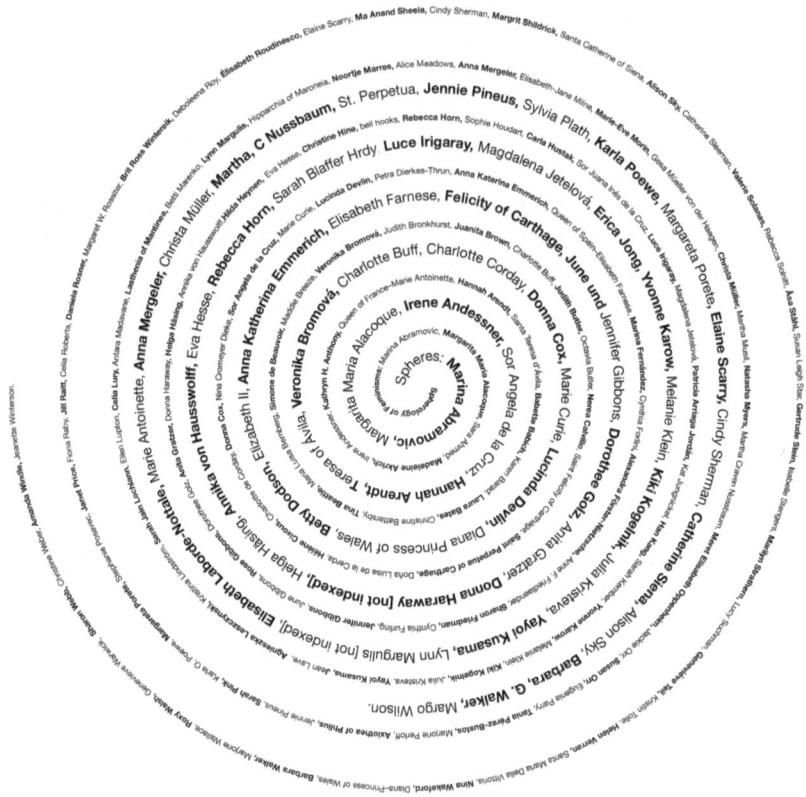

Fig. 4.5 To counter the Spheres index, the spiral of who counts provides at a glance an alternative dance of the indexes to both *Spheres* and *A Companion of Feminisms for Digital Design and Spherology*. Along the shortest spiral of concentric circles are all the women in *Spheres* including those that I'd argue were important enough to include in the *Spheres* index, notably—Haraway and Margulis. The longest spiral shows are all the women that appear in my companion text. In the community spirit of an activist spiral dance, the two indexes pass by one another. I could have turned this sphere upside down to emphasise feminisms rather than *Spheres*, but the reader can rotate the page just the same. Like many ancient spiral petroglyphs, the spiral index can be approached from many perspectives (*Source* Windle)

global topic approached by colonial means (4). This formula describes the minimum complexity you need to arrive at an encyclopedic read of Spheres as world-leading. Designers are involved in this consideration since for them being-*in-the-Spheres-index means dwelling in a book [and] which requires considerable endurance and tolerance on behalf of a discerning reader* (Fig. 4.5).

NOTES

1. Antara Madavane found an error related to June and Jennifer Gibbons in a draft of this chapter while workshopping this book at the University of Linnaeus, Kalmar, Sweden. I'd reversed which twin had died accidentally when editing using an automated grammar-checking software. It's an error as grave as Sloterdijk inferring Caryl was Cheryl Whittier Chessman, discussed later in the chapter.
2. See Butler interviewed by Scott MacLeod for the Cairo Review where Butler regrets not being more active for feminists in her earlier writing (2016).
3. Bear in mind, that it was the secondary literature by Wakeford and Houdart that kept me reading *Spheres.*
4. The indexical notes to this chapter are available here: www.amandawindle.com.
5. An author counts once, even if there are multiple references because my count is for remarks per author and not per reference (unless I state otherwise).
6. *A Thousand Plateaus* is a text briefly cited in *Spheres.* The text by Stengers refers to indexes and the text by Battersby is the same text as discussed in Chapter 2.
7. Judith Butler makes a thoughtful point in an interview with Scott MacLeod, that the author could have done more in her earlier work for those who find gender binaries supportive (2016).
8. The 5.7% average is for all authors from Table 4.1, except I discount Battersby as an "outlier" (an outstanding count from the margins). When including Battersby's interventionist count then the average would rise to 14.2%. For secondary literature in Table 4.2, the counting of female citation rises to 10.7% if considering Raschke and Wakeford, as "outliers". If all authors are included, it's 12.5%. If an index were to become more diverse in general (rather than just "counting on" gender-aware writers like Battersby and Wakeford—for the sake of future researchers) it would still take generations of authors to achieve this and a 1% annual average increase in female index references per year:

$$\frac{5.7\% \text{ (between } 1997-1998) + 10.7\% \text{ (}2008-2013)}{16 \text{ (years)}} = 1.03\% \text{ (per year)}.$$

9. I commissioned Joshua Pitt to write about silenced issues in a blogpost for Backchannels. This blogpost introduced me and other STS scholars to Alice Meadows and the Elsevier report (2017).

10. ORCID is the unique identity code for academic authors.

11. The *New York Times* review was by Oliver Sacks entitled "Bound Together in Fiction and Crime" (1986).

12. As an avid knitter (and sometimes quilter and sewer) myself, there is much to be drawn in knitting to other design disciplines and others have written much more about this elsewhere in feminist technoscience. See Kat Jungnickel for sewing (2018); Daniela Rosner (2018) and Tania Pérez-Bustos for weaving (2016); and Kristina Lindström and Åsa Ståhl for textiles and sewing (2012). These authors are co-related so as to build a cohesive group of references for further research. They appear in a footnote because they do not refer to the main argument of the paragraph. I make this clear so as to make sure a reader does not think I am co-isolating textiles endnotes from digital design references for some other reason.

13. Roderick Stackelberg's book review suggests: "...Yvonne Karow (1997), the author of earlier work on the movement of Bhagwan Shree Rajneesh [Osho] and the Unification Church of San Ayung Mun, applies the insights gained in her study of religious cults to National Socialism" (1999).

14. Twenty-tens refer to the decade 2010–2020. Quantified-selfers are those that measure themselves through various digital (often fitness) devices. Data-visualisers are those that visualise big digital data sets.

15. See Chapter 1 for Sloterdijk's original citation.

References

Ahmed, Sara. *Queer Phenomenology: Orientations, Objects, Others*. Durham: Duke University Press, 2006.

———. *Living a Feminist Life*. Durham: Duke University Press, 2017.

Bates, Laura. 2012–. "Everyday Sexism Twitter Feed." Accessed January 21, 2016. https://twitter.com/EverydaySexism?ref_src=twsrc%5Egoogle%7Ctwcamp%5E-serp%7Ctwgr%5Eauthor.

Battersby, Christine. *The Phenomenal Woman: Feminist Metaphysics and the Patterns of Identity*. Cambridge: Polity Press, 1998.

Cambrosio, Alberto, Peter Keating, and Andrei Mogoutov. "Mapping Collaborative Work and Innovation in Biomedicine: A Computer-Assisted Analysis of Antibody Reagent Workshops." *Social Studies of Science* 34, no. 3 (2004): 325–64.

Cixous, Hélène. "The Laugh of the Medusa." *Signs: Journal of Women in Culture and Society* 1, no. 4 (1976): 875–93.

Couture, Jean-Pierre. *Sloterdijk*. Cambridge: Polity Press, 2016.

Deleuze, Gilles, and Felix Guattari. *A Thousand Plateaus; Capitalism and Schizophrenia*. London: Althone Press, 1988.

Dieter, Michael. "The Virtues of Critical Technical Practices." *Differences: A Journal of Feminist Culture* 25, no. 1 (2014): 216–30.

Elden, Stuart. *Sloterdijk Now*. Edited by Stuart Elden, 1–17. Cambridge: Polity, 2012.

Elsevier. 2017. *Gender in the Global Research Landscape*. Accessed June 24, 2017. https://www.elsevier.com/connect/gender-and-science-resource-center#benchmark-report.

Frampton, Edith. "Fluid Objects: Kleinian Psychoanalytic Theory and Breastfeeding Narratives." *Australian Feminist Studies* 19, no. 45 (2004). http://www.tandfonline.com/toc/cafs20/current.

Gibbons, June. *Pepsi-Cola Addict*. New Horizon: Bognor Regis, 1982.

Häsing, Helga. "Anthology." In *The Feminist Encyclopaedia of German Literature*, edited by Friederike Eigler and Susanne Kord, 20, 1997.

Häsing, Helga, and Ludwig Janus. *Ungewollte Kinder* (Unintended Pregnancy). Berlin: Rowohlt Taschenbuch, 1994.

Hexham, Irving, and Karla O. Poewe. *Understanding Cults and New Religions*. Michigan: Wm. B. Eerdmans, 1986.

————. *New Religions as Global Cultures*. New York: Perseus, 1997.

Houdart, Sophie. "Copying, Cutting and Pasting Social Spheres: Computer Designers' Participation in Architectural Projects." *Science Technology Studies* 21, no. 1 (2008): 47–63.

Iso 999: 1996 Guidelines for the Content, Organization, and Presentation of Indexes. London: British Standards Institution (BSI).

Jungnickel, Kat. *Making Things to Make Sense of Things: DIY as Research and Practice*. London: Goldsmiths University Press, 2018.

Kang, Han. *The Vegetarian*. London: Portobello Books, 2007.

Karow, Yvonne. *Deutsches Opfer - Kultische Selbstausloschung Auf Den Reichsparteitagen Der Nsdap*. Berlin: Akademie Verlag, 1997.

Kerouac, Jack. *On the Road*. London: Penguin, 1957.

Klein, Melanie. "Infantile Anxiety-Situations Reflected in a Work of Art and in the Creative Impulse." *International Journal of Psycho-analysis* 10 (1929): 436–43.

Lindström, Kristina, and Åsa Ståhl. "Working Patches." *Studies in Material Thinking* 7, no. 4 (2012): 1–17. AUT University. https://www.materialthinking.org/sites/default/files/papers/SMT_V7_P4_Lindstr%C3%B6mSt%C3%A5hl_0.pdf.

MacLeod, Scott. 2016. "Global Trouble (Interview with Judith Butler)." *Cairo Review*. Accessed August 8, 2017. https://www.thecairoreview.com/q-a/global-trouble/.

Marres, Noortje. "May the True Victim of Defacement Stand Up! On Reading the Network Configurations of Scandal on the Web." In *Making Things Public*, edited by Bruno Latour and Peter Weibel, 486–89. Cambridge: MIT Press, 2005.

Marres, Noortje, and Richard Rogers. "Recipe for Tracing the Fate of Issues and Their Publics on the Web." In *Making Things Public*, edited by Bruno Latour and Peter Weibel, 922–33. Cambridge: MIT Press, 2005.

Nussbaum, Martha C. *Cultivating Humanity: A Classical Defense of Reform in Liberal Education*. Cambridge: Cambridge University Press, 1997.

———. "Aristotelian Social Democracy." In *Liberalism and the Good*, edited by R. Bruce Douglass, Gerald M. Mara, and Henry S. Richardson, 289. London: Routledge, 1990.

Pérez-Bustos, Tania. "El tejido como conocimiento, el conocimiento como tejido: reflexiones feministas en torno a la agencia de las materialidades" (O tecido como conhecimento, o conhecimento como tecido: reflexões feministas sobre a agência das materialidades) [Weaving as Knowledge, Knowledge as Weaving: Feminist Reflections on the Agency of Materialities]. *Revista Colombiana de Sociología* 39, no. 2 (2016): 163–82. https://revistas.unal.edu.co/index.php/recs/article/view/58970.

Poewe, Karla O. *Universal Male Dominance: An Ethological Illusion*. Amsterdam: Elsevier, 1980.

Raschke, Carl. "Peter Sloterdijk as First Philosopher of Globalization." *Journal for Cultural and Religious Theory* 12, no. 3 (2013): 1–19.

Rosner, Daniela. *Critical Fabulation: Reworking the Methods and Margins of Design*. Cambridge: MIT Press, 2018.

Rossiter, Margaret W. "The Matthew Matilda Effect in Science." *Social Studies of Science* 23, no. 2 (1993): 325–41.

Sacks, Oliver. October 19, 1986. "Bound Together in Fiction and Crime." *New York Times*. Accessed August 5, 2017. http://www.nytimes.com/1986/10/19/books/bound-together-in-fantasy-and-crime.html?pagewanted=all.

Sloterdijk, Peter. *Spheres, Bubbles: Microspherology*. Translated by Wieland Hoban. Vol. I. Los Angeles: Semiotext(e), 1998.

Solnit, Rebecca. *When Men Explain Things to Me*. London: Haymarket Books, 2015.

Stackelberg, Roderick. "Reviewed Work: Deutsches Opfer: Kultische Selbstauslöschung Auf Den Reichsparteitagen Der Nsdap by Yvonne Karow." *Central European History* 32, no. 4 (1999): 493–5.

Stengers, Isabelle. *Power and Invention: Situating Science*. Translated by P. Bains. Minneapolis: Minnesota Press, 1997.

———. "Introductory Notes on an Ecology of Practices." *Cultural Studies Review* 11, no. 1 (2005a): 183–96.

———. "The Cosmopolitical Proposal." In *Making Things Public*, edited by Bruno Latour and Peter Weibel, 994–1004. Cambridge: MIT Press, 2005b.

Teil, Genevieve, and Bruno Latour. 1995. "The Hume Machine: Can Association Networks Do More Than Formal Rules? Constructions of the

Mind." Accessed April 15, 2009. www.stanford.edu/group/SHR/4-2/text/teil-latour.html.

Treister, Suzanne. 2007–2008. "Alchemy Project." Accessed January 21, 2016. www.suzannetreister.net/ALCHEMY/ALCHEMY.html.

Van Tuinen, Sjoerd. "Air Conditioning Spaceship Earth: Peter Sloterdijk's Ethico-aesthetic Paradigm." *Environment and Planning D: Society and Space* 27 (2009): 105–18.

———. "A Thymotic Left? Peter Sloterdijk and the Psychopolitics of Ressentiment." *Symploke* 18, no. 1–2 (2010): 47–64.

Wakeford, Nina. 2011. "Replacing the Network Society with Social Foam: A Revolution for Corporate Ethnography." Accessed August 8, 2017. http://epicpeople.org/wp-content/uploads/2014/09/Wakeford_repla.pdf.

Walker, Barbara. *Women's Encyclopedia of Myths and Secrets*. New York: HarperCollins, 1983.

Wallace, Marjorie. *The Silent Twins*. Edited by Jon (Director) Amiel. British Broadcasting Corporation, 1986.

Webb, Sharon. 2015. "Requirements and National Digital Infrastructures: Digital Preservation in the Humanities". Accessed August 5, 2017. http://sro.sussex.ac.uk/57120/.

CHAPTER 5

A Baroque Performance of the Female in Body Copy and Page Layout

Abstract The design-in of this chapter concerns body copy in relation to the page layouts of *Spheres* and the exhibition catalogue for *Making Things Public* (Sloterdijk in Latour and Weibel 2005). Body copy is the name given when words on the page are visually prepared. Part one considers the mysticism of Santa Teresa d'Ávila and how Sloterdijk uses provocative imagery of female-becoming as performative cuts in textual ideas. Part two examines Peter Sloterdijk's collaborative design work using speculative scenography (2005). By understanding design methods as intellectual know-how, I argue typographically against the credits for this work. These parts are a baroque contribution to the know-how of "doing diagrams" taken from a practitioner's perspective (Verran and Winthereik in Law and Ruppert 2016).

Keywords Baroque · Critical design · Know-how · Mysticism · Speculative design

BEING-IN-A-BOOK

MYSTIC: Reading about religious mystics like Santa Teresa d'Ávila in Sloterdijk's chapter (1515–1582) "Closer to Me than I Am Myself: A Theological Preparation for the Theory of the Shared Inside", makes my body recoil (I, 539–619). It scrunches inwards—not just in a way that furrows the brow and marks the face deep in philosophical thought—but through a much bigger expression that doesn't just belong to the head.

© The Author(s) 2019
A. Windle, *A Companion of Feminisms for Digital Design and Spherology*,
https://doi.org/10.1007/978-3-030-02287-7_5

It's a full-bodied inward hug, an over and under body move that is done while seated at a desk. It's a posture that protects the torso involving the whole body, binding the legs in a crushing spiral (Wirzel), a triple loop of one leg around the other. From thigh to ankle, the body twists towards the designers' desk, plunging downward (Absturz), curving and turning the upper body (Kehre) in a move that cuts creases into *my* chest (Erfahrung).[1] This is the opposite posture of the sculpture that Sloterdijk depicts, which cuts an erotic perspective on mysticism. The Ecstasy of Santa Teresa d'Ávila is by Gian Lorenzo Bernini, an Italian artist and architect working during the Catholic Counter-Reformation.

BEING-IN-A-CATALOGUE

FUTURE-VISIONARY: When I peruse the exhibition catalogue *Making Things Public* (MTP 2005) for Sloterdijk's collaborative work about Pan-European political intervention in remote places, it makes me go still. My body is flat, not in the sense of a tent-like deflation but rather a distributed flatness related to the page. With one palm I press the book flat, and with the other, I draw my finger along running headlines (titles on a page). My body curves inward but not in the way I expressed above. It's about getting my eyes close enough to the page to read the very small citations that run along the margins of the page. This is how I view Sloterdijk's uneven philosophy written *with* and *against* design. The catalogued exhibition is predominantly for an audience of practitioners that study at Zentrum für Kunst und Medientechnologie in Karlsruhe (ZKM). The exhibition brought together collaborators (like Gesa Müeller von der Haegen) and visitors (like myself) from art, architecture, design, sociology, feminist technoscience and STS.

BEING-UNDERGROUND

FIRE: I am editing this chapter on the way to the British Library in London while travelling underground. Male and female travellers sing Champs Elysee in French aboard the tube train heading to Kings Cross. They accompany their song with a plastic electronic saxophone-keyboard, singing because it brings joy and self-esteem to a certain global predicament; plus, it brings in the cash. Seated on this Saturday morning train are nine people, all of which are religiously posturing in public. In front of me, a tall black man is reading something like *How to be Born Again* while another bulkier man with silvery dreads sits in the disabled seat reading *American Gods* (2017). Is this the fiction writer, Colson Whitehead? I hope so. To my left are six women robed in shades of grey and black, chatting to one another. The woman nearest to me elbows me out of the way, pushing past my writing in an attempt to knock it out of my hand. For a moment, *she is closer to me than I am to myself* (I, 539–619). There's anger present in these six teens who also push past the chorus-raising song of these traveller-musicians to exit at Old Street just before Angel.

Unfolding the Diagrammatic Baroque with Theatrical Baroque

Dealing with some of the theological issues within *Bubbles (I)* in the first volume of *Spheres* means re-engaging with some of the damage done through religion, but with that harm in mind, I have written this chapter so as to only include the theological and religious work I felt necessary to contextualise a chapter on the topic of autobiography. It is, therefore, a partial account of religion in *Spheres* because as some queer-feminists already know (i.e. Donna Haraway), this is often a road to nowhere.

My chapter is deliberately structured with many openings and begins disruptively. By way of the three vignettes, I explore what the philosopher, Jacques Derrida titles as the *Margins of Philosophy* and move design and feminist practice from the readerly margins of *Spheres* ([1972] 1982). I also revisit margins as the space surrounding the written word that designers refer to as "body copy". The relation between autobiography and design is a tricky and unexplored juxtaposition to *Spheres*.

When Sloterdijk places images with text, the image can be disruptively performative. This will be baroque to someone who has a Euro-centric art historical education and will know that it's a movement in the seventeenth century where painting, sculpture and architecture intertwined. An STS scholar will know this if reading John Law's sociological chapter in the edited collection entitled, "Modes of Knowing: Resources from the Baroque" (Law 2016). Baroque scenes are often allegorical and the performances are dramatically excessive, unfolding erotic and deadly stories executed in billowy drapery. These artworks are made of the most sumptuous materials of marble, gold and lashings of stucco (plasterwork). Baroque scenes from religious heavens to mythical lands emote in the most shocking of social places: masturbation in the church, orgies in the fountains and rape in the villas of Cardinals and merchant families. The Baroque style was no less sexual or dramatic in writing either, yet since psychoanalysis, the sculpture of Santa Teresa strikes silence in writers and autobiographical censorship. Take for instance, when Law suggests that:

> They [the allegorical stories in baroque] work as mute witnesses to the alternatives that have been written out of the record. This means that it is a task of the baroque scholar to find ways of giving them voice. (Law, 41)[2]

Law asks "what is this baroque 'mode of knowing?'" while offering literature from the history of art for sociological readers such as Genevieve Warwick (2012) and Rudolph Wittkower ([1955] 1997). My focus on Santa Teresa is to understand it as an exemplar of how women's sexualised bodies give *Spheres* its performative register.

My chapter in two parts is baroque in the contemporary sense of the term. With all of this complexity, just to work with the mystics included in Sloterdijk's chapter requires a feminist interdisciplinary enquiry drawing on five disciplines: art history, literary studies, theology, psychoanalysis and sociology. My juxtaposition of two parts takes inspiration from the anthropologists in practice—Helen Verran and Brit Ross Wintereik. Their chapter on baroque modes of knowing brings together a comparison of a tapestry from the sixteenth century and a new wave energy technology research project in Denmark. They call this method a way of "doing diagrams" (2016, 198).

Whose Imagery Counts in *Spheres*?

Some readers of *Spheres* may find the violations towards women could repeat the trauma from which they are trying to heal. What helps my body to recover from the posture described in the first vignette is to do some counting to unwind the baroque loops that stress my body. Counting is soothing when focused on pale saints in the depressive sphere.[3] Of the 49 female references in *Spheres*, 18 are for images rather than text (derived from visual and performance artists). That means 37% of all women indexed in *Spheres* are from the visual sphere. And, the number of women featured whose bodies appear within the images will be significantly higher than the female image-makers. For instance, there are no images accredited to women in *Globes (II)*, the second volume of *Spheres*.

Women, however, appear abundantly in the selected imagery of *Spheres*. It is their traded bodies as pictures that are prevalent—whose ideas and thoughts are always secondary to the philosopher's ideas. When *Spheres* was published, most of the female contributions were from living artists residing in Austria, Germany, North America, Prussia and Sweden but who were originally born in Austria, Czech Republic, Germany, Japan, Prague, Sweden and Yugoslavia (and whose work travelled even further around the globe). To name them all, they are Marina Abramovic (installation and performance artist), Rebecca Horn, (installation and performance artist), Magdalena Jetelová (installation and land artist), Yayoi Kusama (installation artist and writer), Cindy Sherman (photographer and performance artist), Lucinda Devlin (photographer of execution sites), Anita Gratzer (photographer working with Thomas Macho), Annika von Hausswolff (photographer and visual artist), Eva Hesse (sculptor with latex), Kiki Kogelnik (sculptor and painter), Irene Andessner (self-portrait painter), Veronika Bromová (computer manipulation), Donna Cox (pioneer of computer art work), Dorothee Golz (3D design and industrial and architectural design), Alison Sky (3D design and architectural drawing) and Charlotte Buff and Jennie Pineus (whose work cannot be found in searches from the internet or from the British library catalogue).

Unfortunately, publications after *Spheres* are no better. For instance, out of more than 20 references in the chapter "Design" in *The Aesthetic Imperative: Writings on Art*, Sloterdijk might have added a single female reference given his following statement on gender: "A designer cannot regard himself or herself only as a curator of what is already there" ([2014] 2017, 92). My attention to who counts has digressed beyond the discussion of Santa Teresa in *Spheres* and is jumping into the post-*Spheres* writing discussed in part two. To take a step back, I'll return to the importance of religious counting that Law and Sloterdijk use to structure their ideas.

The seven layers of Sloterdijk's microspherology and Law's seven modes of knowing are two examples of religious riddles by the number seven. There are seven sacraments (which in Catholicism includes confirmation) of taking the body of Christ as symbolised in a wafer given by a male priest.[4] Rather than parody their counting riddles, I stay with the rite of confirmation to suggest that the white marble of the sculpture of Santa Teresa represents the brittleness of those egg whites that get separated out and whipped into communion bread and which symbolise the fleshy body of Christ. These flat and brittle wafers get hardened to stone in seventeenth-century Baroque.[5] They are a superfluous ritualistic device to Santa Teresa's nearness to God.

Sloterdijk attends to male and female mystics, but cuts through a diversity of beliefs to place primacy on a white, male dominant theology in Christianity and my own whiteness in some respects frustratingly will add to that account of a "good" story. There are female religious martyrs, mystics and healers (associated with witchcraft). They die in gruesome ways or live extreme and tragic lives centred around a notion of spiritual or carnal love. It's from the biographical details of these women's bodies rather than from their writing that Sloterdijk interlocks the processual ritual of nearness to God to the dimensionality of *Spheres*. To put it crudely, those women removing the authority of patriarchy tend to end up dead, or nearer to it.

There's nothing amusing about Sloterdijk's parody of religious women. It's female tragedy. The nine-dimensionality of Sloterdijk's microspherology can be attributed to nine religious women. They are Margarita Maria Alacoque, noted for her "large cultic movement in post-reformation times" (I, 126); Sor Angela de la Cruz for helping the abandoned poor (III, 860n109), Santa Teresa d'Ávila for her lack of speech (I, 43), Anna Katerina Emmerich—Augustinian mystic and healer that used magnetic treatment of one women less well than herself (ibid., 224), Saint Felicity of Carthage for her turbulent life story (II, 669), Anna Mergeler, a healer acquitted of witchcraft due to her affair with Anton Fugger, of which Sloterdijk describes her as an "attractive healer [...] who had slept with a priest" (ibid., 829); Saint Perpetua of Carthage in relation to Saint Felicity of white women saving women of colour (ibid., 669), Margareta Porete and Santa Teresa d'Ávila are noted for their processual thinking of her being-close-to-God (I, 562–566) and finally Santa Catherine of Siena who is noted for her mystic heart exchange (ibid., 103, 109–12, and 117). To counter this nine-dimensional thinking, I'll name Sor Juana Inés de la Cruz from the New World (Mexico) who called a spade a spade and who found both antagonism and peace in the religious counts of literary chapter and verse. Sor Juana (1651–1695) is a contemporary of Bernini (1598–1680). For Sor Juana, Sloterdijk would not be revered in philosophy as the ace of spades, but as a spade of spades.

Example 1. Countering the intra-action of female performativity: Santa Teresa-Margareta

I've countered the lack of Santa Teresa's own words to be found in *Spheres* by including them below. In the excerpt, Santa Teresa is writing about a vision of a crow gathering around her with a range of (phallic) weapons: daggers, knives, swords, lances and rapiers. She writes:

> I was in great spiritual distress and did not know what to do, when I raised my eyes to the sky and saw Christ – not in Heaven, but far above in in the air – holding out His hand towards me and encouraging me in such a way that I no longer feared all these people, and they could not harm me, try though they might.
> —Santa Teresa d'Ávila, (1515–1582) (1957), 746.4/819

Her words are adjacent to those of Margareta Porete, whose mobility is more important to the *Spheres* argument than what she had to say. From "The Mirror of Simple Souls", the thirteenth-century mystic situates a phenomenological question beyond the death of the body. Porete writes, "Of them that be perished, and in what, of what, and for what . . ." (1927, 124/x). I aim to counter Sloterdijk's impression of mystical women with their theo-autobiographical writing. I also support their words with another saint—Sor Juana—who wrote prolifically from convents in Mexico City. Her insightful quintillas document female oppression in the Baroque form in a less sombre and witty tone. Sor Juana wrote in "reply to a Gentleman from Peru, Who Sent Her Clay Vessels While Suggesting She Would Better Be a Man", in poetic form. Upon being viewed as not feminine, she goes on to retort:

> . . .as I will never be a woman, who may as women serve a man. I know only that my body, Not to either state inclined, is neuter, abstract, guardian, of only what my Soul consigns.
> —Sor Juana Inés de la Cruz (1651–1695) (1997), 137–141

Contemporaneously, they help to contextualise Sloterdijk's juxtaposition of Margareta and Santa Teresa as baroque STS (of doing diagrams).

I do all this because Santa Teresa is another example of how *Spheres* uses the biographies of women. Sloterdijk doesn't cite the Saint's words because they've been rendered performative through Baroque sculpture. He does so to make a performative point but he inadvertently represses the feminine and the female biographical account, which is key to understanding the mystic's impetus to write and speak autonomously, and whose words were written at Toledo, Spain while a guest of Doña Luisa de la Cerda.

MY SITUATED KNOWLEDGE OF BEING-IN-A-CHURCH

For parity on intimacy—so that the reader of this book can get *closer to me than I am to myself*—this section is autobiographical so as to offer a situated reading of Santa Teresa and to practice what I suggest, that is to include autobiography as insightful self-recognition.

QUEERNESS First, Teresa is my Catholic given name but Bernadette (of Lourdes) was my confirmation name and my first queering act. Bernadette reminded me of Bert from Sesame Street and the Saint Bernard dogs who rescue people on mountains with alchohol. I treated my confirmation name a bit like a nickname. My declaration to God was more to do with multispecies intimacy, and the diversity to be found in Sesame Street from the Children's Television Workshop than it was to God. Not that I really knew that at the time, but as I would find out, I had more in common with cities like New York and London than I did with Rome (Santa Teresa's chapel), Lourdes (Bernadette's grotto) or the Lake District (the place where I predominantly grew up).

DISTRIBUTED BODIES So, I had a spiritual education influenced by my two "whole" parents to use Sloterdijk's terminology. While one "whole" of my six (including step-) parents were educated in an English Catholic boarding school and monastery, the other "whole" parent practiced in the Kadampa Buddhist tradition founded by Geshe Kelsang Gyatso. Two partial-parents (step-parents of which there were four) are also Catholic from Sicily and Portugal. Church for me was a place of mediation; by that, I mean where I went to meet one parent away from the other.

SITUATED KNOWLEDGE I re-encountered Santa Teresa again through an art history trip to Rome. I requested (that's the kind of secondary school I went too) to view the sculpture because it was outside of the trip itinerary. As was fitting for an Irish, Jesuit-taught art teacher (and a well-known practice-based historian of cake icing and excellent teacher—Mr Ivan Day), he took me to see such a piece and he did so as a situated learning. So, within the early Sunday morning mass conducted in Italian, I encountered the sculpture.

Within a small church named after Santa Maria Della Vittoria, there to the side of me was Santa Teresa in her chapel. I viewed the sculpture from the pews, glancing in between a congregation of mostly older Italian women who

I stared at intermittently for the length of the service, intercut by gazing at the artworks and embalmed bodies, while partially engaged in the sitting, kneeling and standing of mass. A few of the women I remember well because they sat in the sight line of the priest. They're memorable because of their thick, brown fur coats that were an unfamiliar site in the Lake District at that time.

VISTAS OF ECONOMY What did this sculpture mean to me at that time? Suffice to say, I wanted out of this "being-in". What I noticed was cruelty and wealth within the fur coats and the glittery gold earrings, rings and necklaces worn by the congregation. I can also remember the trick of the light that emanates from the gold rays behind Santa Teresa. A light that's supposed to lead towards heaven but in this Baroque scene leads to more wealth and a lot more voyeurism from the figures that stare down from above. The rays direct the line of flight upwards to the patrons of the sculpture seated in their theatre box. The patrons are carved in bloody marbles, while the salt marble of Santa Teresa (who then and still now) looks shockingly white.[6]

There's a cherub to the side of Santa Teresa. Law describes the relation of two as "the love and affection that a parent might show for a child or the Christian God might feel for his children. At the same time, she is in a place of ecstasy, 'afire', as she puts it, with love" (Law 2016, 24). Cherubs do cheeky things in the Baroque; they pull back curtains to intimate scenes. Like female sexual pleasure. There's a bit of menace in this particular cherub's action who holds the arrow like a writing quill, a directional marker, marking time on Santa Teresa's words.

Let's not forget that in part the viewer is observing a literal as well as a conceptual act of stabbing, be this of self, Godly or sexual love or as an intertwining of all three. Why do writers never refer to the violence depicted in this sculpture? From a situated knowing of reproductive, relational, structural and systemic violence what's coming to the fore is something that gets carved out of religious patronage. Law's description, like Sloterdijk's, is meant to make the reader challenge themselves, but I question their methods as a voyeuristic mode of knowing.

How Santa Teresa's Image Performs
for the Reader of Spheres

The words of Santa Teresa and Margareta are remiss from *Spheres* because Sloterdijk is critical of female mystics for over-dramatising their relation to God. Their emotions seem too much for Sloterdijk to bear witness

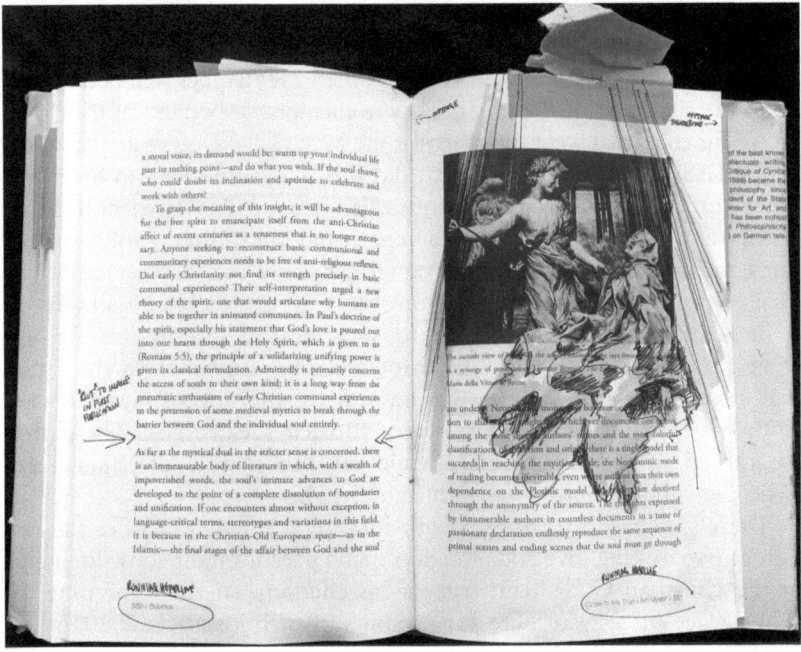

Fig. 5.1 I've drawn into *Spheres* the full figure of Santa Teresa d'Ávila by Gian Lorenzo Bernini (I, 561). Two aspects of this work are missing. The theatre box of patrons situated above and to the side of Santa Teresa, and the elliptical dome that leads up to a lozenge-shaped window created from smaller panes of amber, yellow and white stained-glass from which a radial light emits as if from heaven above (*Source* Semiotext(e)-Sloterdijk-Windle)

to in the written form. The two mystics appear in order alongside five key theological figures; they are Santo Augustinus, Margareta Porete, شهاب‌الدین سهروردی (Shahab al-Din Yahya Suhrawardi), Nicholas of Cusa and, يوحنا يالدمشق or Ἰωάννης ὁ Δαμασκηνός (Saint John of Damascus), but no mystics from India. The two female saints are situated between two lengthier considerations of Santo Augustinus and Saint John of Damascus (four pages to Margareta, half a page to Teresa and more than seventy to Augustinus and John).[7] Three out of around eighty pages relate to Koranic theology rather than to Catholicism, yet Sloterdijk argues that whether he takes his source from "Christian-old European space" or "Islam", he says that the final stages between God and the soul are for the passionate female mystic: "...the same sequence of primal scenes and ending scenes that the soul must go through on its way back to One" (I, 561–2).

My own autobiographical story is also revealing, but I purposefully removed the excitable speech from my story to remain *a little closer to myself than I am to the reader*. In other words a bit of privacy and non-disclosure. Sloterdijk dramatises Santa Teresa in image only, that of Fig. 5.1. Sloterdijk places the image of Santa Teresa on the right-hand side of a double-page spread, midway through a text about Margareta Porete. It's not quite the same place in the German-Suhrkamp (1998) and English-Semiotext(e) (2011) publications because the Semiotext(e) edition is a larger book. It's clear cut in the original German edition because it's a combined page and paragraph break beginning after the paragraph cited below. The American-English translation splits the paragraph into two further down this same paragraph. Interestingly, the text seems to excuse the lack of words cited by Sloterdijk, for either mystic (Suhrkamp I, 573).

> As far as the mystical dual in the stricter sense is concerned, there is an extensive body of literature in which, with a wealth of impoverished words, the soul's intimate advances to God are developed to the point of a complete dissolution of boundaries and unification.
>
> —Semiotext(e) I, 560[8]

How Were the Mystics Chosen?

What are the alternative modes of knowing that Sloterdijk could have made? Preceding the discussions on Margareta and Santa Teresa, the text focuses on Santo Augustinus as a mystic, but not in the traditional sense (I, 547–8). This isn't a new idea. For instance, Petra Dierkes-Thrun, a literary scholar, notes that Oscar Wilde referred to Salome as a mystic in a letter to the artist and stage designer Charles Ricketts (Dierkes-Thrun 2015). He argues that Sloterdijk's choice of mystics is arbitrary but that's not the case: Santa Teresa is the "money shot" to a sort of centrefold to the chapter (1998, 561).

Sloterdijk could have cut the chapter with an image of Santo Augustinus (by the same sculptor) but Bernini had less freedom of expression with this commission. The bronze cast is placed beneath the seat at the altar of Santo Peters in Vatican City. Sloterdijk could have used this to extend and make clear his single-sentence point that Fatherhood is in crisis or acknowledged the Christian Fatherhood's repressive techniques towards both men and women during the Baroque period (I, 641). He could have drawn on his contemporary experiences with the Indian mystic, Osho, to enable the reader access to what is offstage in Sloterdijk's autobiographical trilogy, but perhaps this brings the reader far too close— *closer to Sloterdijk than he is to himself* which is a matter of *doing disclosure*.

So why does Sloterdijk choose Santa Teresa? He considers the temporality of her move towards God (in all the female Christian examples) as the most sped up of all, with John of Damascus and Santo Augustinus the most measured. The body of the radical mystic (Santa Teresa) annihilates the individual self at the moment when a nearness to God becomes configured as a being-in-God. I'd argue next that Sloterdijk is more than ready to "write out" rather than keep possessively "writing in" allegories about female sexual liberation without acknowledging the cost to the individuals. Sloterdijk stands authoritatively and goads his readers to risk more than he will ever risk himself. The reader is always *closer to themselves than they are to Peter Sloterdijk*.[9]

With Margareta Porete Against Santa Teresa d'Ávila

We know of Marguerite Porete that she sometimes travelled through the country like a show-woman, reciting from her mirror of souls in front of highly diverse audiences. The Neoplatonic diva managed to prove her contemporaries that the enjoyment of God — which was simultaneously the first legitimised form of self-enjoyment — can liberate itself from church walls and churchmen; Marguerite Porete is one of the mystical mothers of liberality. Would this make mysticism be the matrix for performance art? Would performance then be the impulse that releases the subject? Would the subject be the manifest side of bi-une movedness? Would movedness be an emergence from the shared? And God an expressionist through the woman?

—I, 566

Without evidence, Sloterdijk relies on the contested idea that Porete travelled the country to communicate her doctrine. Sloterdijk makes this amplification to stress a performative point on mysticism and further this with the image placement of Santa Teresa.[10] Sloterdijk's reasoning around Margareta's impetus for mysticism may have been an act of survival but one that ultimately led to self-sacrifice. The writers that contest the gendered depictions of Santa Teresa and Margareta Porete are from English literature (Petra Dierkes-Thrun), theology (Tina Beattie) and the history of psychoanalysis (Élisabeth Roudinesco), rather than art, architecture, design, philosophy, sociology or STS.

Beattie, a Professor of Catholic Studies, wrote about Porete in *The Guardian* as "A Forgotten Female Voice" (2010). Beattie notes that Porete posed a threat to male-dominated ecclesiastical hierarchies because she was removing a particular patriarchal control and Christian in-hood over her faith with God. In relation to Santa Teresa's words, Sloterdijk replaces them with her image and by doing so, he encapsulates the body of the Saint within his own "phallic economy" (Roudinesco 2015). Sloterdijk never refers to Bernini's sculpture in the prose of his trilogy but he does pay attention to Ignatius Loyola whose Society of Jesus (which became the Jesuits) was an order for priests. Santa Teresa founded the separate Discalced Carmelite order for nuns which was sanctioned by Loyola in 1593 (see Jill Raitt in Mehl 1998, 213). It was one of the most punitive orders of the Catholic church. See the life history of the Latin-American-bisexual-lesbian-Baroque-nun, Sor Juana Inés, who belonged to both the Carmelite and Hieronymite (a more creative and lenient) order, for a comparison of the two. Sor Juana Inés de la Cruz's poetry, writing and philosophy were published for male and female patrons of the Spanish viceregal court in Mexico.[11] Sor Juana is one of the most notable female exclusions in Sloterdijk's argument, which could help to contextualise the struggles of Santa Teresa and Margareta Porete by telling a different story about women living a mystical life.

I'd argue that both Sloterdijk (1998) and Law (2016) are attempting to write in their religious modes of knowing with the sculpture of Santa Teresa (as I've done), but in Law's case he veers away from autobiography: "We may know through love, but unlike Serres, we do not talk about it. Or we may know through spiritual experience (Santa Teresa), but then, like Karl Jung, we hide this reality too" (Law 2016, 19). There's a lack of vulnerability in both writers' account of Santa Teresa, which leaves me to question the cuts, and to an extent, ask biographical questions about Law and Sloterdijk. Disappointingly, neither contextualises the risks of feminist autobiography to make a point that cuts in writing, lead to wounds of the flesh. Whether Law or Sloterdijk write this way because they've been disciplined to write confidently or because they simply had beneficial heteropatriarchal lives, is spurious biographical speculation hardly worth the mention. What the viewer can do is self-recognise because their modes of knowing are not the only ... *Theories of the Shared Inside.*

The first part of Sloterdijk's chapter title parodies that of Santa Teresa's autobiography, *The Life of Saint Teresa of Ávila by Herself* ([c.1536] 1957). Sloterdijk repeats an act of exclusion by cutting Santa Teresa's words. His placement of Santa Teresa's image within *Spheres* is a quip that goes something like this: *If Santa Teresa needed no mediator for her communion with God, then Sloterdijk need not include the words written down by Santa Teresa because she cares less about the men that define religion.* Beyond religion, what's needed in an account of the mystics is to remember the pain suffered by some for writing their feelings down. Unlike Santa Teresa who was revered as a saint after her death by illness, Porete was burned at the stake for circulating her book in Old French rather than in Latin.

In *Spheres* the sculpture of Santa Teresa amplifies Margareta's so-called "diva" persona (I, 566). Pre-Baroque paintings of Margareta (1250–1310) and Santa Teresa (1515–1582) are serene yet restrained portraits of feminised piety. Being-in-contemplation-with-God is depicted as a painterly stillness. Sloterdijk doesn't explain this; he *just* acts baroquely by placing Santa Teresa's "coming"—to use Jacques Lacan's expression—in between his passages on Santo Augustinus and Margareta Porete. This cut gives Margareta's travelling persona a sexualised edge. Law and Sloterdijk remain too much in the voyeurism of it all and their enquiries similarly echo the psychoanalyst Lacan: "What is she [Santa Teresa] getting off on? It is clear that the essential testimony of the mystics consists in saying that they experience it, but know nothing about it" ([1998] 1972–1973). Sloterdijk appreciates women who bare their flesh, be that a female activist, a nude model or a robed saint.[12] Flesh is what Sloterdijk gets off on. To an extent, his trilogy of *Spheres* is a tale of his writerly addictions to sex and violence towards women, tenuously veiled in a backdrop of late twentieth century, feminist sexual liberation. He is, in a sense, writing both *with* and *against* the sexual liberation of women of which Santa Teresa is the exemplar performance.

ATTENDING TO THE PALENESS OF SANTA TERESA D'ÁVILA

As I said earlier, there's an entanglement of love going on in the image of Santa Teresa—of a stabbing that produces no blood. Like so much heteropatriarchal violence, is it all in one's head? I think not. At that time, Santa Teresa's emotive embodied writing could have been a risky act leading to potential self-harm and a death sentence. Consider the pale pallor of Santa Teresa as the contemporaneous paleness of all women with anaemia, whose blood regardless of skin colour, is stretched elliptical—whose blood is not always whole and when viewed under a microscope is visibly broken. Both science and religion tell us this mystical woman is almost dead. When Bernini renders Santa Teresa's words voyeuristically as a scene after sexual release (his or hers?), is he mocking the female orgasm or just laughing at what he can get away with exhibiting in a church?

By getting to know oneself at the time of the Baroque was—let's not forget—a matter for many female saints of life and death. To me, this sculpture depicts the act of loving the female-self despite theo-heteropatriarchal containers of being-in-the-world. Also, Bernini's sculpture embodies the very tension between a voyeuristic-hetero-patriarchal-sexualisation which can be further, and a priori, considered by multiple modes of sexual knowing: masturbation (mono), bisexuality (non-binary), gay and lesbian love (queer or binary), trans-encounters (distributed) and pan-gendered knowing (many).

What Law and Sloterdijk (like Lacan) get stuck on when looking closely at Santa Teresa are descriptions of the individual body parts that can be seen because she's mostly draped in cloth. This way of seeing or mode of knowing (especially when reading *Spheres* twenty years after it was first published) reads more like contemporary descriptions of body dysmorphia. This is especially so when Sloterdijk describes heart exchange and menstruation blood in his tenth Excursus; this is more of an uncomfortable read than the chapter on mysticism, from which it follows (I, 619–24).

The bodily detail, however, is crucial to Sloterdijk's description of microspherology. In *Spheres* I, the interfacial is centre stage in the microspherology, but what about hands? Sloterdijk considers the "inter-genital relationship" and the "inter-manual connection" to be like coitus and the handshake (to be peripheral to his microspheric analysis), and yet with Teresa it's pivotal (I, 541). For Law, the cherub's hand (writing) and Santa Teresa's feet (floating) get a descriptive account too. Sloterdijk downplays Porete's death and plays up to the sexualisation of Santa Teresa's ecstasy, albeit peripherally. Law downplays Santa Teresa's writing to over-emphasise the entanglement of Church and parenthood.

The sculpture's stoned-whiteness symbolises a mobility beyond the physical containers of theo-heteropatriarchy. The ability for Margareta to travel is the element that Sloterdijk focuses on; however, by tracing the work of Sor Juana, his account could have taken a further journey to find an alternative female viewpoint of Baroque in Mexico City and a joyful-quipping opinion at that.

Example 2. Guessing About Gesa Müeller von der Haegen

Almost a decade after the German publication of *Spheres*, Sloterdijk collaborated with the Belgian architect and scenographer Gesa Müeller von der Haegen in the exhibition, *Making Things Public* (MTP). While this wasn't the most intriguing of exhibits, it's important because it's Sloterdijk's only collaboration with designers. I next ponder Müeller von der Haegen's contribution as a scenographer and what the practice of spatial performance design offers to a philosopher like Sloterdijk? This shift to practice (post-*Spheres*) afforded Sloterdijk another performative mode of knowing with a practitioner audience.

RUNNING HEADS AND KNOW-HOW

The following analysis and continuation of work with diagrams is further inspired by the essays of American designers Ellen Lupton and Abbott Miller (1996). They make semiotic statements by including icons instead of words; moreover, typefaces and font sizes vary to emphasise an argumentative point about design. I do so to research power in the entanglement of methods from design and philosophical ideas.

Broadly, the catalogue is a public record that brings together art, design, philosophy and sociology, as well as fine art practice. I'll show how a running head (or headline) is more than just a title (in Sloterdijk's two pieces) because they are, par exemplar, a way to acknowledge individuals in writing. A running headline is a section of a page that is assigned for titles and usually the domain of the graphic designer to standardise. They can appear at the top, bottom or to the side of a page. The running heads now up for discussion can be found in the catalogue that accompanied the exhibition (which contains two essays) and are a critical juncture between the writers, editors and designers of the catalogue. Design *methods* can be observed as secondary to political and philosophical *ideas* evident in the running heads, which is the reverse of Lupton and Miller's essay-method.

The first piece of writing in the catalogue is Sloterdijk's standalone philosophical essay and the second is a collaborative design work. Both have different styles of running heads that indicate to the reader when they are within the design of the catalogue or in the second kynical (a more forceful mode of cynicism) "advertising" sphere. The first essay by Sloterdijk has the same running heads that all sections of the catalogue have, which are designed by Holger Jost and Christine Weber. However, the second, entitled "Advertising", differs because its running head appears at the top of the page rather than to the side. This has been designed by Sloterdijk's collaborators (most likely Mark Jongen and Gesa Müeller von der Haegen).

Sloterdijk and his collaborators are engaged in making a political parody through design practice and that takes know-how. Know-how is a legal and methodological word used in R&D, indicating what a practitioner learns, brings to and takes away from a collaborative project. Know-how is related to intellectual property and is, I'd argue, the most important yet salient aspect of the law that design practitioners should know about at this present time (i.e. forget patents). *Making Things Public* places high value on interdisciplinary collaboration *with* specialists in philosophy and design. Through a range of design exactitudes, Müeller von der Haegen and others provide the philosopher (Sloterdijk) with design "know-how": namely, the application of design standards as a visual method of cynical critique. Additionally, these collaborators also have the know-how to work with a philosopher as a client-collaborator.

In the introduction to the catalogue, Latour refers to Sloterdijk's written contribution as a fable (MTP, 17). While storytelling can be a method of design, I'd argue these methods make up more than half of the story about the catalogue's record of collaboration. Design counts for more in part two than what is given credit for because there are ideas in methods of which I'll explain. Figure 5.2 shows what an interactive designer might call a "wireframe" or what an art historian might call "figures of composition", but these can also be referred to in advertising and print-design as a "scamp" or "mock-up". These methods relate to different areas of communicating design ideas and they differ in their modes of knowing.

Sloterdijk's design collaboration involved a team from two institutions made up of five named collaborators. The composite credit for collaboration is misleading. Captions to images are the place to give credit to the collaborators, but in a sense, they are a vista into what happens offstage. The credits frame the content of the catalogue's page.[13]

To an extent the three-dimensional visualisations showing political activity within an emergency scenario work better within the catalogue

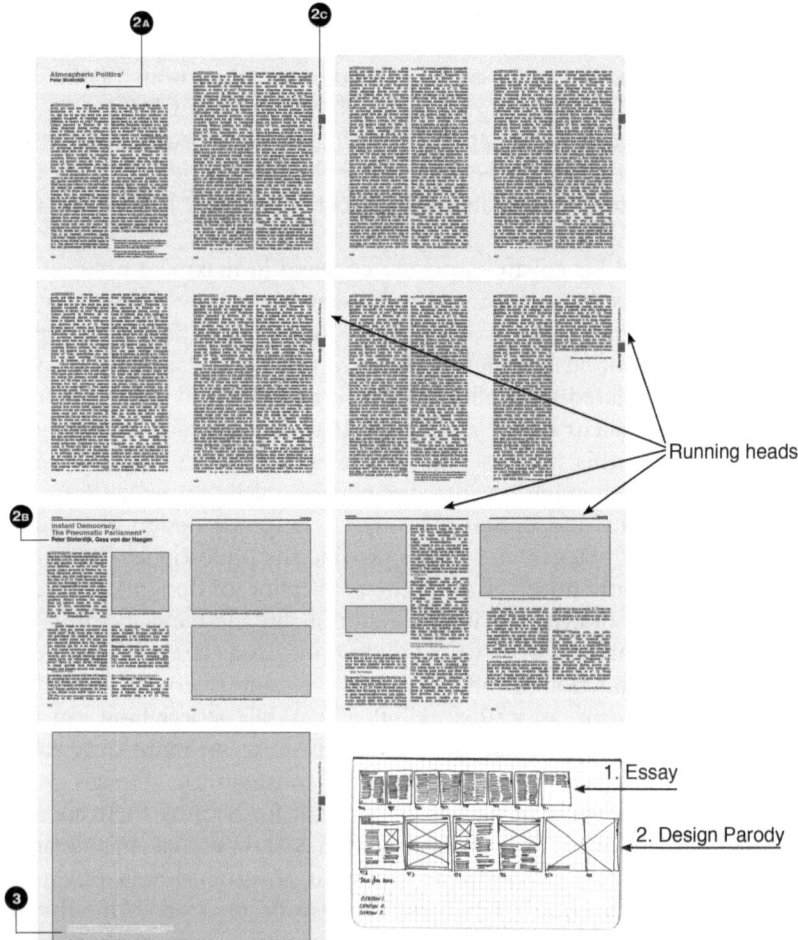

Fig. 5.2 Digital and hand-rendered mock-ups of Sloterdijk's contributions to the *Making Things Public* catalogue. The circular numbers show the range of crediting formats (*Source* Author's own)

than they did as an exhibit to be seen in situ on account of its flatness and its relation to the composition of a double-page spread. The design spreads depict a critique of Pan-European political intervention as international terrorism situated in Libya, Iran, Cuba, North Korea, Syria and Sudan (MTP, 952). Within the landscape scene, there is a "pneumatic" parliament. The structure looks new and watery, and fresh and airy in contrast to the arid surroundings. Air-dropped from above, the inflatable bubble-shaped dome tent is juxtaposed with a fortress pictured in the distance that's built from the rock of the surrounding area. The vista encapsulates an architectural imaginary which makes a kynical point that parliaments can work in abstraction when democracy and policy operate at a distance.

There's much need to question what design is doing in politics today. When I participated as an arts and culture expert in a "Voices of Culture" (VoC) for audience development via digital means (which is a structured dialogue for informing and potentially contributing to making policy) with the European Commission in Brussels, part of this process of reporting back to the Commission included participating in a world café (Saldanha et al. 2015; Windle 2016). It's Juanita Brown and David Isaacs's method of knowledge exchange derived from business and economics which helps to enact change (2005). As I witnessed it, the method jutted-up against the interior design. The policymaking environment with its hard desks and immovable furniture (which are excellent for presentations) were not ideal for initiating a world café scenario and so it spilled out into the corridors and other non-spaces of the building where round tables rather than oblong ones could be found.

Sloterdijk's collaborative essay with Gesa Müeller von der Haegen is a critique of how political ideas are made, where they are situated and how far removed contemporary politics can be from the places where political decisions get made. However, to make a more contemporary comment on design methods in politics, Sloterdijk wouldn't have had to go very far to see it in action within the rooms of the Commission when the world café methods were "airdropped" into a room of immovable furniture and the dialogue broke out of report-like modes of dissemination. Sloterdijk could then have made a more realistic observation (closer to home) on how far design methods intervene in today's politics of ideas. There's a point of contradiction that can be made by bringing together an observation from both the catalogue pages of *Making Things Public* and my anecdote from *Voices of Culture*. If designers are afforded political action tantamount to international terrorism (as exhibited in MTP), then the role of arts and cultural organisations (VoC) differs for it is meant to provide cohesion across countries and to pick up the pieces (and broken people made homeless) through the way design has the capacity to negligibly intervene in foreign policy.

Using Designerly Exactitudes as Critical Design

Staying with the design exactitudes to be found in the catalogue pages, I'll consider next the precision of European and international standards parodied in Sloterdijk's collaboration. Note the use of DIN standards (Deutsches Institut für Normung—the German Institute for Standardization).

> The degree of reflection of the opaque areas of the inlay achieves a value of 80 percent according to DIN 5036-3 . . . DIN 5036-3 for instance is the measurement for light — it's transmission & reflection of photometric properties: — "Range of Available Models of the Pneumatic Parliament – Color-psychological Specification [...] This enables a nation-specific, color-psychological and color-coordinated nuance to flank the democratization process".
>
> —MTP, 954–55

The catalogue does not follow national paper-sizing standards, not even the Colombian, North American, Japanese or broader formats like Imperial and Duke. This means that the catalogue is expensive and non-standardised in size. This print run required specially cut paper creating a lot of excess paper in the off-cut. However, my point is not about environmentalism, recycling or carbon footprints, but to say that the catalogue design has not undermined the design argument contained in the mock-project. For this exactitude, credits are due to the catalogue's designers—Jost and Weber.

There are standard ways to credit in a catalogue for an exhibition, yet in the juxtaposition of the two essays by Sloterdijk, the crediting gets complicated. Are the credited contributions actual or fictional? Are citation credits a place for yet more parody? My questions are somewhat baroque. There are five places in the catalogue and three differing collaborative citation methods. The first citations occur in the contents pages which are mirrored in the bylines (see Fig. 5.2, n2a–2c): the first essay credits "Sloterdijk" and the second essay with collaborators credits "Sloterdijk and Müeller von der Haegen" (MTP, 944 and 952, respectively). The third citation occurs at the end of the second essay as a full double-page spread over the top of the large photograph of the exhibition display (see Fig. 5.2, n3). The unique "advertising" running head is no longer used in the final pages. The following credits sit over the top of the image.

Layout 1.

THE PNEUMATIC PARLIAMENT, idea: Peter Sloterdijk, architecture and scenography: Gesa Müeller von der

Haegen, Dierk Jordan, interview: Marc Jongen, 3-D visualization, model, video interview, folder, poster, pro-

duc-tion: ZKM in collaboration with Staatliche Hochschule für Gestaltung Karlsruhe, photo: Franz Wamhof

Layout 2.
THE PNEUMATIC PARLIAMENT

- **idea**: Peter Sloterdijk
- **architecture and scenography**: Gesa Müeller von der Haegen, Dierk Jordan
- **interview**: Marc Jongen
- **3-D visualization, model, video interview, folder, poster, production**: ZKM in collaboration with Staatliche Hochschule für Gestaltung Karlsruhe
- **photo**: Franz Wamhof

Here are two simplified typographical layouts to show the legibility issue of the credits in Fig. 5.2, n3 (in case it was difficult to follow in the preceding paragraph). The first layout is how the text appears in MTP, as a single paragraph in Helvetica (MTP, 956–7). In the second layout, I've tried to make the text more legible. What's revealed is a hierarchy of collaborative practice. Sloterdijk's collaboration is neither symmetrical nor does it need to be, but if Sloterdijk is credited with ideas, surely Müeller von der Haegen could be partially credited with methods? Müeller von der Haegen is not merely artistically rendering Sloterdijk's ideas (or rendering them in three-dimensions) because the know-how of methods are at stake.

The reason for focusing on Müeller von der Haegen rather than the other male collaborators is because the credits above appear as a hierarchical list (and not alphabetically). Sloterdijk is credited with ideas rather than philosophy. Müeller von der Haegen is jointly credited with "architecture and scenography" rather than methods. The cynical design layout follows a tradition of speculative thinking in architecture that was published in magazine and exhibition form, reminiscent of the pamphlets of Archigram in the 1960s by the architects Peter Cook, Warren

Chalk, Dennis Crompton, David Greene, Ron Herron and Michael Webb, along with the designer Theo Crosby. This method is echoed again in the 1990s by Rem Koolhaas, an architect of contradictions and the Professor of Architecture and Urban Design at Harvard University's Graduate School of Design. It's unsurprising to find Sloterdijk presenting on *Spheres*, four years after the exhibition at ZKM. This may have been brokered through introductions from his practitioner collaborators or perhaps it was simply a shared incentive to discuss critical design with a like-minded colleague.[14] While this is speculation, what can be known from the chronology of Sloterdijk's publications is that the initial kynical ideas from a CCR ([1983] 1987) led to the cynicism of *Spheres* (1998), which then got advanced through critical design practice in MTP (2005) and led to further presentations for designers at Delft and Harvard universities (in 2009).

Critical design is an umbrella term for speculative design and design noir which are "an attack on the prevailing culture of design" (Koskinen et al. 2011, 116). Both speculative and critical design aims to help rehearse issues in the future, as play does for children (op. cit., 46). Yet, I'm purposefully selecting a definition that defines critical design in violent terms to emphasise a disjuncture in the purpose of play and political critique made *through* design. Whether it's playful humour, irony, kynicism or English wit, whatever the mode of cynicism, humour in design can be deeply problematic, unreflexive and not *entirely* critical in and of itself.[15] To extend Koskinen's point below, these playful modes of knowing don't always help prepare designers going into industry or academia. Collaborative design doesn't help designers to necessarily collaborate in a critically reflexive and self-recognising way.

> … unless these narrative methods are grounded in real data, they easily reflect only the wants and preferences of researchers. At worst, they become just devices of persuasion".
>
> —Koskinen, op. cit., 114

DUE CREDIT

The catalogue has two further citation frames: the index and biographies (MTP, 1051–57 and 1058–71, respectively). Müeller von der Haegen is absent from the biographies of artists, activists, sociologists and philosophers. Only in the design frame beneath the advertising running heads is the architect/scenographer rendered visible. Below is Müeller von der Haegen's missing biography which was made public six years later for the exhibition "Container Architecture" in Düsseldorf.

> Müeller von der Haegen is an "architect, scenographer and specialist for temporary architecture and urban economics".
> —Müeller von der Haegen in Andreas Rossmann, 2011

At last, Müeller von der Haegen is given full credit, defined in her own words (note the speech marks)—a task Sloterdijk is never burdened with doing himself because his introductions are all too plentiful and heroic.[16] The point I'm labouring is that Müeller von der Haegen's role could have been more clearly defined. Had she understood her "know-how" in the sphere of academic research then a further contribution could have been made: this would be to argue that scenography provides the methods and not just the aesthetic veneer to Sloterdijk's critique. The collaborators provide the ability to parody advertising processes and techniques through a critical-speculative design "methodology" (to couch it in research terms because it's through design as research that contributing credits become more rigorously inspected, or at least that's my hope). Even if collaborative work was done separately, but together, there was more than enough room on the page to have done something else with the credits. My point being that philosophers can learn from designers and vice versa, if only there were discussions about know-how.

Did Sloterdijk's philosophy change as a result of collaborating with designers? Sloterdijk's cynicism-kynicism got more precise by way of employing someone else's design exactitudes, but perhaps this know-how was not enough for Sloterdijk to be convinced that there are ideas in methods which could have been jointly credited in his collaborative work with designers like Müeller von der Haegen. It was not enough to make Sloterdijk rethink what it symbolically means to frontload a single-authored essay, distanced and separate from the space of design parody. Why then did he need so much more space on the page?

Sloterdijk thought less of design for its intervention into politics, but in the process, he thought less of the methods and know-how of the scenographer. Did Gesa not "get her shit together" in time (to repeat again Sloterdijk in, CRR, 89)? Did the ideas not matter? Or, was there a reason why she didn't have her biography together at the time that this catalogue went to press? It doesn't matter (as Sloterdijk would say) because to reply is to be drawn into the game of pointing out a reduction in a woman's words, in the body copy of a catalogue where her biography did not belong. Did Sloterdijk choose not to chase or had Gesa already left the building? Am I trying to save Gesa when she doesn't want, need or care about the "know-how" and the distribution of artistic license? This section was not written for Gesa but rather as a defence of the value of methods as ideas, as a mode of knowing-how to empower design collaborators. In summary, designers ought to defend their know-how, especially when designing with academic and artistic companions so that they know what to do when they aren't indexed and are instead called a muse, a diva or a cunt; because, as I can state for myself, it's important to understand your know-how when in-design, when asked to get your shit together by collaborators like Sloterdijk that write *with* and *against* the canons of a philosophy through practice.

CONCLUSION

Will future writers look back on the future-visioning that Sloterdijk and others made public and approach it from as many viewpoints that Bernini's sculpture of Santa Teresa still evokes today? From what sidelines might these new readings take place? Service designers working in Syria? Terrorists in Belgium? Will future designers ponder the patronage of Sloterdijk's design work or continue to analyse *women as merely architectural units* of collaborative design-making without any consideration for the collaborators' know-how?

Bernini's work draws believers and non-believers in and its intimacy is blatantly seductive. The sculpture will resonate for many years to come because there are alternative readings to be contemplated when considering Santa Teresa as a binary or fluid exemplar of gender-freedom and sexual release. In relation to the second example of my chapter, Sloterdijk's future-scape depicts how Europe assembles in architectural form a mode of decision-making tantamount to potential war (international terrorism) and that's a long-lasting statement to make which remains meaningful a decade on (particularly when pondering the third vignette at the start of this chapter). However, the creditation practices of the collaborative speculative design became a case in point of critical self-recognition for future methods-ideation.

NOTES

1. These phenomenological terms are set out by Marie-Eve Morin but without any relation to gendered bodies (SN, 85–6).
2. What's missing from this chapter is Law's elucidation on Quakerism. To read this is to go back to his co-written text with Michel Callon (2003, 8).
3. Like David Foster Wallace's fiction entitled the *Pale King*, I can empathise with the mind-numbing, yet transcendental ecstasy of counting and accountancy (2011).
4. The count of seven in Sloterdijk's chapter could in fact not be about Christianity but Rajneesheeism. Is it about the rising of the energy known as kundalini or "life force" as the coming together of the male and female as one? As the seven chakras are ascended, it unites in the many-petals of the cerebrum. Could the discussion about female mystics in Catholicism be a cryptic-contemplation on the figure of a specific priestess? Is Sheela (blamed by Osho for the failure of Rajneeshpuram) the priestess of "Jesusgrove", Rajneeshpuram? A Kali of sorts, to draw on the Hindu and Buddhist goddess associated with temporal situations—and nearness to death?
5. Alternatively, this chapter could have echoed the writing of Sor Juana and used the structure of the *quintilla*, the Spanish craft of writing poetry, in a rhyming scheme written in 5-line stanzas. It's a form of poetry that is more about what cannot be done rather than what can be done in words.
6. Bear in mind that Santa Teresa did call-out heretics such as Lutherans and there is intolerance in her words (Cohen, 32.3/819).
7. In the Suhrkamp publication they are referred to as Aurelius Augustinus, Margareta Porete, سهروردی شهاب‌الدین—Shihaboddin Yahya Sohravardi, Nikolaus Cusanus and Johannes von Damaskus.
8. Law also talks about this as undoing boundaries and "formlessness" or "boundlessness" (2016, 31 and 21).
9. He chose to write about Porete and Santo Augustinius rather than the Indian mystic Osho and his assistant Ma Anand Sheela. This shows a trend that what's really central to Sloterdijk's ideas is just out of reach of the reader (if they don't check beyond the given references). Again, is the subject of his jokes really the target of his humour?
10. Santa Teresa's autobiography suggests that there are four processional stages to being-in-God ([c.1536] 1957). There are less steps towards God than the processual stages of prayer for Santo Augustinus. On the ninth day of prayer, Santo Augustinus is revered as a saint and the nature of the problem the prayer-maker seeks is confessed and leads towards a final closing prayer. The nine-dimension of Sloterdijk's microspherology echoes the nine confessional prayers of Santo Augustinus, which in

Spheres, plays out as Sloterdijk's confessional writing-out about the bodies of the nine previously mentioned religious women.

11. For a different account of female writing in the Baroque, see the Argentinian and Mexican films by the Director, María Luisa Bemberg, "I, the Worst of All" (1990) and the televised mini-series created by Patricia Arriaga Jordán (Canal Once 2016) entitled, "Juana Inés".

12. Joel-Peter Witkin's photograph of a nude Venus on the cross is of an unnamed female model who worked at a Lido in Paris. Sloterdijk could have chosen to use another nude crucifixion scene by Witkin, that of "Penitente" (Witkin 1982) a work depicting a gagged Christ which explores New Mexican imitations of Christ as a memorial to Jewish Holocaust victims (Parry 2001, 36). Instead, Sloterdijk chose the other photograph and abbreviates the title, which I'll set out in full: "[History of the White World] Venus Preferred to Christ". This should have led to Sor Juana.

13. Playing with what's to the side of central stage is something the recently discussed Baroque sculptor Bernini understood when depicting the patrons of Santa Teresa who appear just above her looking down from their seats in a theatre box. To an extent, it's what a running head looks like in the Baroque when being-seen in a theatre box replaces the convention of a donors' plaque as a more apt way to recognise sponsors in theatrical Baroque. This is how to carve in masonry the symbolic notion of male patronage as an architectural unit of theatrical financial transaction. I'm parodying here Sloterdijk's phrase of women as an architectural unit discussed at length in chapter three (from TTM, 3).

14. For further doubts on the work of Rem Koolhaas, see Koskinen et al. (2011, 101). The work of the Belgian Professor Hilde Heynen is highlighted as she researches power in the theories of architecture (2003, 43).

15. For a definition of wit, see the critical designer of industrial design Fiona Raby's interview with the late architect Gerrard O'Carroll (2012, 84–5).

16. By Couture (2016), Elden (2012), Latour (2009, 2005), Raschke (2013) and Tuinen (2009, 2010). See Sascha Rashof for more on the limitations of Sloterdijk's *Spheres* theory in her doctoral thesis on makerspaces (2016).

REFERENCES

d'Ávila, Santa Teresa. *The Life of St. Teresa of Avila by Herself.* Translated by J.M. Cohen. London: Penguin Books, [c.1536] 1957.

Bemberg, María Luisa. (Director). *Yo, La Peor de Todas* (I, the Worst of All). Argentina: Crisalida Films, 1990.

Brown, Juanita, and David Isaacs. *The World Cafe: Shaping Our Futures Through Conversations that Matter.* San Francisco: Barrett-Koehler Publishers, 2005.

Callon, Michel, and John Law. October 26, 2003. "On Qualculation, Agency and Otherness." Centre for Science Studies. Accessed January 29, 2018.

http://www.lancaster.ac.uk/fass/resources/sociology-online-papers/papers/callon-law-qualculation-agency-otherness.pdf.

Couture, Jean-Pierre. *Sloterdijk*. Cambridge: Polity Press, 2016.

Derrida, Jacques. *Margins of Philosophy*. Translated by Alan Bass. *Différence*. Chicago: Chicago University Press, 1972.

Dierkes-Thrun, Petra. "The Ecstatic Moment: Mysticism and Individualism in the Ecstasy of St. Teresa and Salomé." Accessed November 19, 2015. https://wildedecadents.wordpress.com/2012/12/17/the-ecstatic-moment-mysticism-and-individualism-in-the-ecstasy-of-st-teresa-and-salome/.

Elden, Stuart. *Sloterdijk Now*. Edited by Stuart Elden, 1–17. Cambridge: Polity, 2012.

Foster Wallace, David. *Pale King*. London: Penguin, 2011.

Heynen, Hilde. "Intervention in the Relations of Production, or Sublimation of Contradictions? On Commitment Then and Now." In *New Commitment: In Architecture, Art and Design* (Reflect 01). Rotterdam: NAi Publishers, 38–47, 2003.

Inés de la Cruz, Sor Juana. *Poems, Protest, and a Dream*. Translated by Margaret Sayers Peden. Introduction by Ilan Stevens. London: Penguin Classics, 1997.

Jordán, Patricia Arriaga. (Director). *Juana Inés*. Mexico: Canal Once, 2016.

Koskinen, Ilpo, John Zimmerman, Thomas Binder, Johan Redström, and Stephan Wensveen. *Design Research Through Practice: From the Lab, Field, and Showroom*. Burlington, MA: Morgan Kaufmann (Elsevier), 2011.

Lacan, Jacques. *The Seminar of Jacques Lacan 1972–1973*. Translated by Bruce Fink. Edited by Jacques-Alain Miller. Vol. XX, *On Feminine Sexuality. The Limits of Love and Knowledge* (Encore). London: W. W. Norton, 1998.

Latour, Bruno. "Spheres and Networks: Two Ways to Reinterpret Globalization." *Harvard Design Magazine* 30 (2009): 138–44.

Latour, Bruno, and Peter Weibel, eds. "From Realpolitik to Dingpolitik Or How to Make Things Public", In *Making Things Public*, 40–61. Cambridge: MIT Press, 2005.

Law, John. "Modes of Knowing: Resources from the Baroque." In *Modes of Knowing*, edited by John Law and Evelyn Ruppert, 17–58. Mattering Press. https://www.matteringpress.org/, 2016.

Lupton, Ellen, J., and Abbott Miller. *Design, Writing, Research: Writing on Graphic Design*. London: Phaidon, 1996.

Parry, Eugenia. *Joel-Peter Witkins. Catalogue 55*. London: Phaidon Press, 2001.

Porete, Margareta [Or Unknown French Mystic of the Thirteenth Century]. *The Mirror of Simple Souls*. Translated by Clare Kirchberger. London: Burns Oates and Washbourne Ltd., 1927.

Raby, Fiona. "Interview." In *Darkitecture: Learning Architecture for the Twenty-First Century*, edited by Gerrard O'Carroll, 84–87. London: Two Little Boys, 2012.

Raitt, Jill. "Laudem Caroli: Renaissance and Reformation Studies for Charles G. Nauert: Sixteenth Century Essays and Studies," Xlix no. 5. In *The Two Spiritual Directors of Women in the Sixteenth Century: St. Ignatius Loyola and*

St. Teresa of Avila, edited by James V. Mehl. Missouri: Thomas Jefferson University Press, 1998.

Raschke, Carl. "Peter Sloterdijk as First Philosopher of Globalization." *Journal for Cultural and Religious Theory* 12, no. 3 (2013): 1–19.

Rashof, Sascha. *Designing Place—Topologies of Maker Labs*. Doctoral thesis, Goldsmiths, University of London [Thesis], 2016.

Rossmann, Andreas. June 30, 2011. "The Hamburger of Architecture." *Style Park*. Accessed August 30, 2015. www.stylepark.com/en/news/the-hamburger-of-architecture/322093.

Roudinesco, Élisabeth. 2015. "The Sculptural Iconography of Feminine Jouissance: Lacan's Reading of Bernini's Saint Teresa." Accessed January 27, 2016. www.lacanschool.org/events/the-sculptural-iconography-of-feminine-jouissance-lacans-reading-of-berninis-saint-teresa/.

Saldanha, Charlotte, Dominic Smith, and Amanda Windle. 2015. "Audience Development Via Digital Means. Voice of Culture." *Structured Dialogue Between the Arts and Culture Sector and the EU Commission*. Accessed October 15, 2018. http://ualresearchonline.arts.ac.uk/8666/.

Sloterdijk, Peter. *Critique of Cynical Reason*. Translated by Michael Eldred. Edited by M. Eldred. Minneapolis: University of Minnesota Press, [1983] 1987.

———. *Spheres, Bubbles: Microspherology*. Translated by Wieland Hoban. Vol. I, Los Angeles: Semiotext(e), 1998.

———. "Atmospheric Politics." In *Making Things Public: Atmospheres of Democracy*, edited by Bruno Latour and Peter Weibel, 944–57. Cambridge: MIT Press, 2005.

———. *The Aesthetic Imperative: Writings on Art*. Translated by Karen Margolis. Edited by Peter Weibel. Cambridge: Polity, [2014] 2017.

Van Tuinen, Sjoerd. "Air Conditioning Spaceship Earth: Peter Sloterdijk's Ethico-Aesthetic Paradigm." *Environment and Planning D: Society and Space* 27 (2009): 105–18.

———. "A Thymotic Left?: Peter Sloterdijk and the Psychopolitics of Ressentiment." *Symploke* 18, no. 1–2 (2010): 47–64.

Verran, Helen, and Brit Ross Winthereik. "Innovation with Words and Visuals: A Baroque Sensibility." In *Modes of Knowing*, edited by John Law and Evelyn Ruppert, 197–223. Mattering Press, 2016. https://www.matteringpress.org/.

Warwick, G. *Bernini: Art as Theatre*. New Haven, CT, and London: Yale University Press, 2012.

Windle, A. June 23, 2016. "A Personal Plea from Europe's Polar Bears" In *New Statesman Tech*. http://tech.newstatesman.com/enterprise-it/european-digital-research-funding.

Wittkower, R. *Bernini: The Sculptor of the Roman Baroque*. London and New York: Phaidon. ([1955] 1997) In *Art and Architecture in Italy 1600–1750: II High Baroque*, edited by J. Connors and J. Montague, rev. New Haven, CT, and London: Yale University Press, [1958] 1999.

CHAPTER 6

How to Design Spherically as a Matter of Recursion

Abstract The design-in of this chapter relates to a one-year interface design. The live, geospatial (GIS) web-based application enables environmental scientists to add, assess and analyse threat assessments to endangered species on a digital map. This requires consensual working on matters of multispecies extinction. Following a testing error involving a cartographic marker, what unfolds is an account of doing atmosphere design. I revisit the geometries of *Spheres* and research into recursive searching and being-in-a-design-issue.

Keywords Atmosphere design · Foams · GIS · Interface · Maps Species · Threat

With a shift towards empirical research, this chapter explores the design and development of a web-interface over the course of a year. The web application was based on a prototype that was already two years into development but which had no interface design. The tool enables environmental scientists to add, assess and analyse threats to endangered species on a digital map. This mapping helps to evidence a threat through written explanations, photography, video and web links. A threat can be the contamination of a food source in a river, the atmospheric changes that bleach coral in a reef, the introduction of a predatory species to a remote island and the release of gases into the air by forest plants or methane ponds. In one place there can be many threats and endangered

© The Author(s) 2019
A. Windle, *A Companion of Feminisms for Digital Design and Spherology*,
https://doi.org/10.1007/978-3-030-02287-7_6

species. Mapping them to coordinates requires drawing skills and discussion through workshop assessments. The project was a partnership
between a global technology company and an environmental organisation. It's a particular kind of design work done at the edge of what
may be possible, programmatically speaking. Results are rendered in the
geographic information system (GIS) and this needs to be done as fast
as possible. A GIS, in this case, is a geospatial web application that is
designed for the internet which connects to a database, and in that sense,
updates "live" or "on the fly".

Finding an error involving two polygons and their markers (see
Fig. 6.1) made it clear that GIS can be as spatially and temporally fragile
as the environmental ecosystems themselves. To consider these systems
as anything else would be an oversight. The impact of the spatial error
was of small consequence to the overall design because errors like these
get smoothed out during testing and demoing. Beta-testing is when a

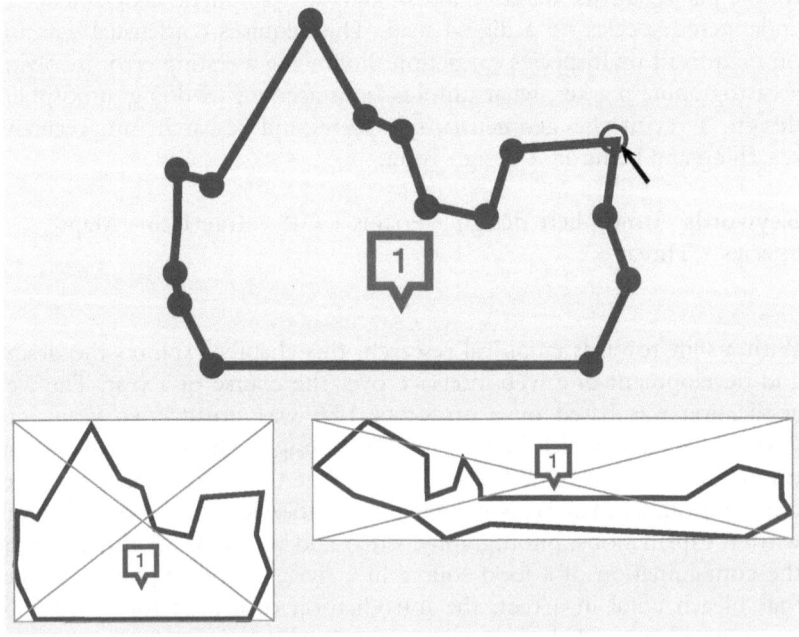

Fig. 6.1 A cartographic marker appears when a polygon shape file is completely
drawn point-to-point (*Source* Windle)

software contains some known and unknown bugging issues. Demoing is when the points of design innovation are promoted. In this application, the main innovation was the ability to recursively search.

This chapter is written recursively with multiple cuts. In fact, it's written closer to how I experienced the project—as a set of ever-changing, collaborative decisions that constitute doing design. At times the chapter oscillates between recounting the polygon-marker error within the design and the embodied situation of doing design. The reader can circle back to recursively read the numbered sections out of order like 3, 4, 5a–b, 1, 2 or 1, 6, 5a–b, 3, 2, 4. By immersing the design reader in a complex global issue the text will weave in and out of solving errors while thinking philosophically, all the while trying to stay on schedule. It's related to *Spheres* through atmosphere design because working with threats to endangered species and the atmosphere designers involves designers, workshop assessors and environmental scientists. The tool is thus firmly located in the depressive sphere.

SEARCH

"Oh, I've submitted it. Let's see what comes out of here?" ... but to be able to do this easily ... to be able to see the land cover behind it and to see things like ... you know ... dams behind it ... then to be able to use that information to draw it ... and then see instantly what the impact is on their species.

—*Rashmi*[1]

One of the roles of the [redacted] project is to find something in space and find protected areas that overlay ... or do some analysis of protected areas based on the spatial ... I want to start doing predictive models of threatened species then I would look to that portal to give me an overview, first of all, of what has been recorded. It is rare that I would want to do it for any one species. It's possible; I might want to look at species of snowdrops and things like that. I do research on the conservation of snowdrops in Georgia ... in Turkey ... and that sort of stuff ... snowdrops are close to my heart [laughs]. But more likely, what I want to do is do it for groups of species.

—*Matthew*

... species distribution is not static and sometimes people just don't know where they are ... the only knowledge is roughly where they are, but you can never pinpoint, say this is the boundary that this species can be found. I think it's, you know, not practical ... But in practice, since we are dealing with

*thousands and thousands of species, it's not always feasible and also the resolu-
tion of the threat maps, they are not as accurate as you know something like the
exacted or the very specific boundary between … you know … the threat areas
and sometimes people just draw just by looking at the map and draw roughly
where the threats are because these "refer maps"; they aren't accurate anyway.*
—Jing-Ru

It's a summer afternoon in June 2012. Jason and I are crammed into a
private office that looks onto ponds and neat oblong fields.[2] We're inter-
viewing Rashmi and Jing-Ru at an environmental NGO in office spaces
crammed with darkly-felted screen dividers and plan chests. Rashmi is
tucked beneath the open rafters of the house, sat at her desk where she
currently works. Rashmi is "responsible for everything related to spatial
informatics starting from data management to applications developments,
to tools, to technical support, to training and to anything related to GIS
capacity". Rashmi will soon move to a new organisation that pays more
as her digital skills are in high demand. Jing-Ru is transient in other ways,
working across teams gaining work experience as an intern. We discuss the
project in a meeting room with Jing-Ru in the same converted farmhouse
that we interview Rashmi.

Matthew Smith was interviewed amongst desktops, tablets and dual-screen
monitors with glass walls full of handwritten formulas in Cambridge,
the United Kingdom. Surrounding the desk is an array of neatly stacked
papers. As a "research scientist mostly working with models of things",
Matthew works in a computational lab for Microsoft Research Ltd. The
air-conditioned environment could be anywhere in America, Asia or
Europe. My workspace is similar except my bookshelves are full of port-
folio-sized design books and texts in gender, media and STS. Rather than
models and formulas, my walls display data visualisations of the DigiLab
and several portable foam boards for mapping and drawing purposes.
Jason works in a shared office situated in a converted warehouse in South
London. His shelves are also full of books, Pantone guides, paper and
printers and coffee-making facilities provided for collaborative working.

To echo Jing-Ru, this is a practical design problem, but it's "not practical". It's philosophical, post-phenomenological even, because the language used to describe the design problem is not *just* about geometry
but about broken world-views, intangible vistas and the impossible idea
of embracing the globe as a complete digital project. The processes of

using a GIS application reveals how important spatial spherological log-ics are in this tool, but rather than ask how *Spheres* can help an R&D (research and development) researcher design more thoughtful cooper-ative, collaborative and consensual digital tools, the question is reversed. How can GIS design revisit the geometric language abundant in *Spheres*? I spent a ridiculously long time (too long to admit) trying to answer the former question but all the while the design practice and my reflection on philosophy were intertwined. This wasn't a requirement for the pro-ject or relevant for sharing with the entire team, and I'm not advocat-ing for this thinking to be a universal way for project leads like myself to reflect on design issues. Pondering an interface design as an atmosphere design is a helpful way to describe the unique spatial entanglements of this interface. As a consequence of reflecting when designing, I was able to make various feminist spherological interventions like attuning to the well-being of the design team and to how our bodies needed to adjust to designing across global time zones.

> *Maybe they [other interviewees] have a different interpretation but what is really valued is the ability to share all of that [moves a cursor across a digital prototype of the GIS] so that you can go "okay, working from that here's the bits I believe; here's the bits I don't believe; here's the bits we need to change to be species specific" and whenever people do that it is always very much appreci-ated, partly because of the difficulty of getting funding to set those things up, partly because of so far a lack of incentive in doing that. It's not something academics can do; they don't get rewarded for doing. It all falls into this gap because also people need to build these things.*
>
> *—Matthew*

Matthew articulates a gap in the labour that academic researchers can produce when building a web application. This leads to questioning what worlds are co-produced by Matthew, Rashmi or Jing-Ru? And, how is data shared among groups of mammologists, freshwater species experts and policymakers?

FROM THE TROPES OF *SPHERES* TO CUTTING ACROSS SPHERICAL LOGICS OF RECURSION

During the initial research, the scientists demonstrated how to search environmental data which became a user-journey of recursion. Sifting and scoping-out live spherical problems involving vectors is essential.

There's a taxonomic search for species and threats and a user can select "species to threats" or "threats to species", recursively refining their selection. There's a reliance on zooming in and out of the map to view shapes, rather than to nets of relational data so to consider the interface as an actor-network is only part of explaining the socio-material relations. It's not a choice of whether the metaphors of *Spheres* or the metaphors of networks work best because they're enmeshed, mixed-up and nearly always combined more so than the literature on networks and *Spheres* suggests (Sloterdijk, TTM; Latour 2009).

I'll explain *bubble issues* and *foam problems* related to the testing of the web application. A bubble-foam enquiry is to ask what it's like to be in an atmosphere designer. The vocabulary from *Spheres*—like "bubbles", "foam", "hyper-orbs", "vistability", "island making" and "atmosphere designer"—vary in helpfulness. The rest of this section will briefly revisit these terms because some are new to understanding atmosphere design.

Foam designs include the labour of creating "facilities, atmosphere, [and] encompassing situations" or "islands" (ibid., 65). Foam designs vary from geodesic domes through to spaceships, individual weapons through to systems of state execution with cyanide gas in *Spheres*.[3] Sloterdijk's idea of islands as a "being-towards-death" is supported by Vilém Flusser's philosophical notion of "anthro-topology" (III, 458 and 522). Designing an interface is a form of *island making* but it's a matter of self-recognition, reflection and retrospection that warrants whether the designer is an atmosphere designer or not.

In the first volume of *Spheres*, the sky is discussed as a "spatialised immune system", in relation to the astronomer and mathematician Copernicus. In later volumes, telecommunications replace the sky as the all-encompassing motif of immunity (I, 25). Confusingly, the spatialised immune system is any "primary around oneself" be that "nature, cosmos, creation, situation, culture, home or dwelling and so on" (III, 181 and 500). These primaries are not the same thing as immune systems that are a "bio-medical articulation of antibodies" but are "static defence measures" against injury (ibid., 213). Sloterdijk's definition of spatialised immunity is a concept embedded in systems thinking but problematic to fully understand (ibid., 213).

Like immune systems, "hyper-orbs" too have transitional meanings in *Spheres*. Hyper-orbs mean "world-soul" when related to the philosophy of Aristotle (I, 69) but they rapidly develop to mean "virtual globalisation" (I, 70–1). This term connects Sloterdijk's philosophy to media references, especially with the cultural theorist Marshall McLuhan for his ideas about the "electronic global village" and how the house is no longer considered the definitive container—as the ultimate extension of the human body (I, 625, and III, 503). Sloterdijk repeatedly criticises the philosopher and political scientist, Gilles Deleuze and Felix Guattari. He comments on how they failed to capture the "orb of being" in their ideas about folds, lines, rhizomes and networks (I, 90 and 137; and II, 41). To consider the GIS interface as a "hyper-orb" would mean comprehending all the data of the web application as a true representation of threats to species. The design team often worked remotely to one another in smaller groups across the globe. In a sense, the situatedness of designing could be described spatially as a hyper-orb, though the pre-fix "hype" seems overly performative.

"Vistability" or vista-viewing, however, are helpful terms to make sense of the relations between species and threats when viewing the data. This is an adaption of the original meaning in *Spheres* which relates to medieval painting. Religious figures are depicted interlocked in an intimate gaze represented by a halo that frames the two figure heads (I, 31). Though Sloterdijk doesn't say so, this form of visual communion was believed to transmit the substance of a thing, transmitted through the eye. Needless to say, this interlocking concept doesn't go far enough to explore today's interactional issues, to consider the interlocking of geometric shapes not as a human-centric doubling, but as ways of viewing visual errors in the GIS.[4]

The roles of "atmosphere designer" (III, 162) and "climate designer" (II, 435), of which Sloterdijk calls "whole-based relationships", are interchangeable terms. They differ from bubble (doubling) relations (like the shared halo just mentioned). Again, atmosphere and climate change design are used in relation to a myriad of professions like "media technologists", "interior architects" and "labo[u]r medics" (III, 65). Atmosphere designers are also considered alongside "climate wardens"— terms he uses to denote "lords" and "wardens" respectively (ibid., 162). Atmosphere designers may well be feminist spherologists, which differs from *Spheres*. Working with what the past century designed into and out of the world-globe requires the premise to engage in embodied matters which, in this chapter, means interweaving a spherology of feminisms into ways of doing design.

A GIS that involves entanglements of multispecies (human, animal and algorithm) soon become increasingly complicated by the various

cuts that environmental scientists make in the data. These cuts prevalently reappear as a series of containment matters in the interface designs
(Strathern 1996). Drawing on Marilyn Strathern's anthropological paper
entitled, "Cutting the Network", I'll explicate on cutting the spherical
logics of recursion rather than networks (ibid., 521). Reading *Spheres*
while doing design also became entangled in issues residing in the testing
processes. While testing is an opportunity for teams to make changes,
they are somewhat limited in scope at the beta-phase and decisions need
to be made as to which issues get priority.

Design Methods—Working with Spatial Intra-active Errors

While there are user-experience techniques which can be used to consider intra-active errors, these problems can get situated between various types of testing methods undertaken by project managers, designers,
developers and users. My role was to lead (as Co-Principle Investigator)
the research and the design development team from end-to-end which
is a role of translation and mediation. I led the interview research and
regularly liaised with my co-lead at Microsoft Research Ltd. We often
worked to different time zones (Seoul to Seattle) and during the project the interface design team, the tech company and the conservation
organisation were physically located in other parts of Britain, Denmark,
Germany, India, Italy, North America, South Africa, South Korea and
Switzerland. For the partners, this often meant working against our bio-
rhythms (sleep) and redefining our working days to complement the
general well-being of the distributed team, i.e. making sure meetings
weren't too late for childcare or too early for team members to be awake.

Personas, user-journeys, task models, design and competitive reviews
were created as part of the interface design methods. Some user-journeys were created by the conservation organisation and the technology
partner but were revised as a result of the research with twelve environmental scientists and GIS experts through semi-structured interviews and
workshop discussions. I worked closely with designer, Jason Rainbird
and developer Richard Mason, to integrate the frontend of the interface
with the backend of the database. The design team expanded to more
people when we needed to talk out the complexity or required further
skills designated to a particular issue. To write this chapter, I referred
to my journal notes from all meetings, as well as emails, hand-rendered
and digital sketches, wireframes, key textual and numeric documents like

status reports, technical specification documents, Gantt charts and work plans, sprint updates and product backlogs. I could suggest that in terms of project management methods, I worked "agile" (teams work simultaneously on multiple problems), somewhat in "scrum" (devising problems and reiterating them together in small phases called "sprints") and hardly at all in "waterfall" (working on separate tasks or problems individually, one person tasked at a time). I'd say that I worked conceptually together, and most of the time philosophically, alone.

This sort of computational design research is rare in R&D but there are exemplars that do not promote shortcutting this additional work academic-oriented designers do, like that of Alex Taylor et al. in "Modelling Biology—Working Through (In)Stabilities and Frictions" (2014). While drawing predominantly on the STS work of Karen Barad and Helen Verran, they interweave ideas by John Law. To summarise a key point I'd like to repeat:

> ...This places the onus on scholars not only to produce accounts of practice, but to also accept their active participation in them. [concluding that] Design at the interface, in this sense, becomes one of the ways in which we as science scholars might "nudge programs of research" (Law 2004, 40), be they in biology or otherwise.
>
> —Taylor et al. (2014)

The Logics of Points and Polygons

Some governing premises and material practices were kept in mind when designing the GIS interface. There are four sets of Cartesian logics: (1) the programmed algorithms of the backend database, (2) the programmed frontend interface, (3) the representational geometries utilised in the visual design, and (4) the standardised numeric taxonomies that identify a species, threat and intervention ([1921] 1922). Additionally, this database of spatial shapes is one of the largest, global vector-based databases. Our searches meant it had to computationally "return" results at speed—preferably in seconds rather than minutes, hours or days. Therefore, all spatial issues were governed by temporal concerns. Speed was the presiding computational limitation. This GIS was designed to help localised threat assessments. At all times we endeavoured to create an interface that held this premise above all logics of computation.

Circles, Spheres and Freehand Drawing Tools

At the beginning of this chapter, I explained roughly what counts as a threat and what counts as a species, but now I'll deepen those definitions. Threats to species are events like logging and volcanoes, but threats also contain what are called interventions which are actions like introducing an invasive species or creating a national park. Additionally, a threat can consist of multiple definitions like volcano and ozone chosen from a list of taxonomies (see Fig. 6.3). The shape files contain listed information about endangered species—that is plants, mammals, birds and amphibians in both their Latin and common names (i.e. Leo and Lion).

Each shape has a marker, a numerical system to help locate and view individual and groups of shape files. A threat "area" can be any geometry (except a rectangle or square) drawn with a mouse (see Fig. 6.1). A threat assessor can go into the preloaded data and pinpoint a specific threat or draw a rough shape of the impacted area (see Figs. 6.1 and 6.2). Accuracy varies depending on what is being drawn and the assessor's chosen method, which is agreed by consensus. For instance, mammologists draw irregular polygons, whereas those mapping freshwater species draw a succession of small wiggly lines (like linear ripples showing the direction of movement up or downstream). I was told by Edward, a workshop assessor, that the freshwater mapping was the most consistent and sophisticated consensual type of hand-drawn methods among the species assessors, scientists and volunteers. Workshops bring together experts with regional insights into threat-mapping to share their maps of threats to endangered species. By placing these on a digital map the data can be shared across continents with regional and global significance.

Base-Layers and Map Projections

Various mapping projections can change by selecting different base layers. The Peters and Mercator maps are two projections that show the world flattened in different ways and depict parts of the globe bigger (i.e. stretched) than other parts (i.e. landmass near the equator is shown less significant than the two poles), whereas the Gall-Peters projection is rectangular with no north-to-south distortion. Base layers can show data on a light background, but it can also be shown in reverse.

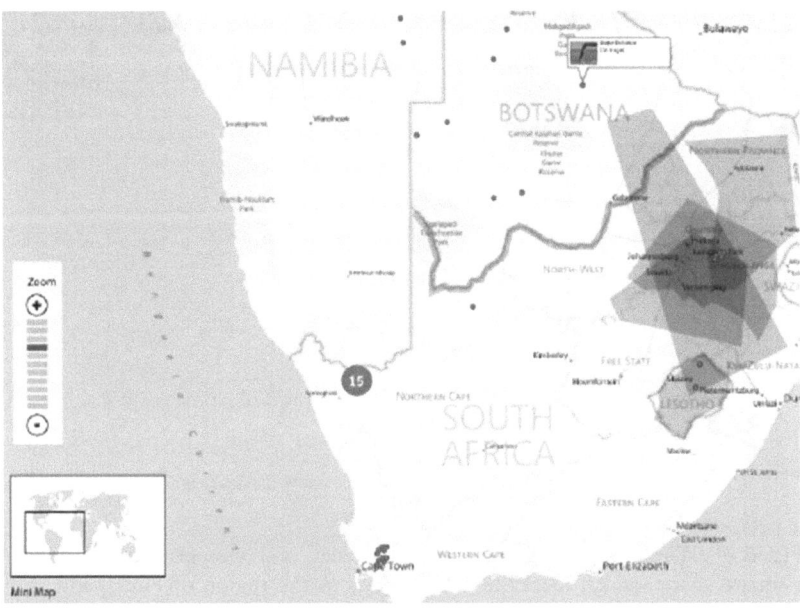

Fig. 6.2 Wireframe shows threats drawn as a point, line and polygon onto a base layer map. It's a fictional example to illustrate the range of drawing tools to be developed. Figures 6.2, 6.3, and 6.4 show only bing™ maps as base-layers (*Source* Microsoft Research Ltd.)

Figure 6.4 shows a threat polygon. The coloured shape has a solid westerly edge drawn along an existing line on the map's base layer. The easterly edge, however, is less jagged or exact and doesn't seem to follow a physical feature on the base layer like a river or a mountain range. Assessors can view these shape files on different base layers; for example, maps that show the braiding on rivers are more apt for mapping freshwater things.

Wireframes like Fig. 6.4 are the designer's blueprint which helps developers to build the interface. In a wireframe, the features and functions are all shown, but the styling and design details are paired down. Brand guidelines were introduced when the design became more finalised. The issues affecting the design and development were predominantly interactive rather than about styling issues to do with brand or issues related to base layers.

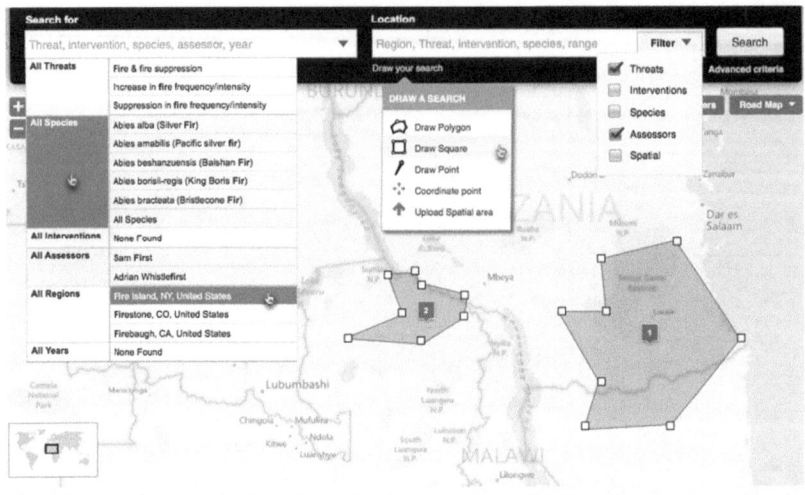

Fig. 6.3 Wireframe shows several shape files and the process of drawing a square shape, which identifies a Fir species in the region of Fire Island, New York. This isn't part of a user-scenario, but simply illustrates some of the tools and palettes (*Source* Microsoft Research Ltd.)

Cartographic Markers and Transparency

In this section, I highlight the clarity of the polygon-marker issue and the error occurring in the testing phase. The marker that appears in the middle of a spatial shape appeared in early wireframes because it's an essential way-finding icon found on both printed and digitised maps. A window shows the polygon's content. Figure 6.4 (Sect. "Base-Layers and Map Projections") shows a threat in the Kabo Forest Concession: the threat is "hunting and trapping". This content pane tags the spatial range to an endangered species. A marker is therefore not a tag.

It wasn't possible for markers to be placed in the centre of any shape as intended, but there were other ways of making shape files distinct from one another. The various aspects of a polygon were broken into smaller visual problems, namely colour, marker, transparency and zoom. The polygons could be identified by size and colour coding, but when overlapping, they become indistinguishable (see Fig. 6.4). The spectral range was, therefore, a limited way to differentiate vector-based data. Setting transparency (by an interactive toggle) extended this range

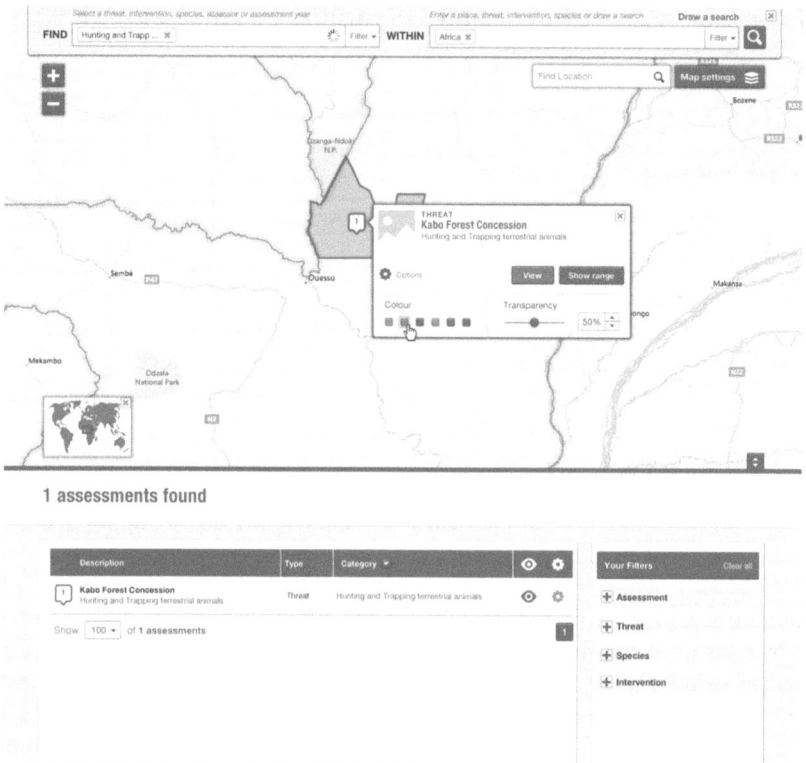

Fig. 6.4 The wireframe shows the search function for finding an existing threat in the Kabo Forest Concession (*Source* Microsoft Research Ltd.)

by helping to distinguish shapes that overlap. Layered shapes can be moved to the surface or to the bottom near the map interface by clicking on an "eye" icon in the table of contents next to the list of species taxonomies.

Workshop assessors search and sort 1000 species in five days, so identifying species quickly in the interface is another temporal priority (i.e. volunteers' time is valued by other means than economy). The process requires expert knowledge and cooperation to reach a consensus on which shape files will remain as evidence. A marker was meant to help differentiate polygons at speed. An environmental scientist using the application may need to identify information by sifting through all the data, as Lucas Joppa explains below.

> *In the example of returning [searching] an entire species like frogs is [sic]*
> *that you get 25,000 species overlapping, you don't want that, so you search for*
> *a genus of frog "hyla".*
>
> *—Lucas, July 2012*

Lucas is a computational and environmental scientist and reptile expert and he conceived this project, subsequently winning prestigious awards for it since I worked (remotely) with Lucas. We were the two Co-Principle Investigators of this interface design development. We remained in regular contact sometimes updating each other weekly, daily or even hourly. I gained further discourse, help and support especially from Kristin Tolle (in Seattle, America) and Kenji Takeda (in Cambridge, Britain). Their bridging roles in the discourse around intellectual property and planning were some of the most thoughtful and supportive I've experienced in R&D.

Knowing what data to search for is preferable because complex combinations of data will appear. Challenges involved both visibility (a relational issue of viewing one polygon beneath or next to another, nearby or far away) and vistability issues (the relational issue of zooming in and out of various mapping scales). This web application's interface opens up the complexities of species-mapping. Lucas is saying that searching more generic levels of species will return a large dataset which is not ideal because it's a case of trial and error. That's why searching recursively is so important as a process of refinement.

> *Mapping, editing and searching isn't a linear set of steps. A user might want to*
> *edit then go back and repeat searches to keep adding threats to multiple species.*
>
> *—Aiden, June 2012*

Three weeks were spent solving the recursive search problem which meant when problems like the polygon/marker issue finally emerged, it was competing against other priorities. From the outset, there were other constraints to finding the polygon problem including partial data in the backend database which affected the returned results.

To recap, markers were in the interface from the outset and were the primary solution for viewing and identifying overlapping polygons.

Figures 6.2, 6.3, and 6.4 (Sects. "The Logics of Points and Polygons", "Circles, Spheres and Freehand Drawing Tools", and "Base-Layers and Map Projections"), show the designers working almost consistently from a local viewpoint. The polygon-marker problem became most apparent when viewing different continents at the same time by zooming out to a worldview. Scaling out from a local view is crucial in threat-mapping because endangered species are not always present in one continent (like the Panthera species). Furthermore, not all species are immobile. Species move between continents, across national borders and conservation boundaries. Risks too are changeable: like a volcano eruption near a reserve area could be a threat to an entire ecosystem depending on where the lava flows. The threat of a diamond mind polluting a river will be a threat to those drinking from it rather than living within it. Illegal hunting and trapping may also affect more than one species of mammal in the same vicinity where they are prized for their skin and where bone poachers frequent. When threats to humans get entangled with threats to other species, these become evident too: such as in the notable identification of illegal hunting for bushmeat at a time when food is scarce for all. The spatial mapping of an endangered species and a threat are temporally related and both require constant updating in regular workshop assessments.

Recursive Searching

I wondered how we hadn't resolved the issue of plotting a marker in the middle of a shape file? If markers were not part of the wayfinding system, how and to whom does it matter? The markers sunk to the bottom of a digital ocean. The error stretched two shapes across an ocean which led to the cartographic marker being dropped. To be precise, it went to the bottom of a Kanban—a list of "Do, Doing and Done" tasks—and then was later dropped altogether. I traced the design issue as a processual timeline to find out how and when the error emerged and when the differences between an antelope and a Panthera species mattered. Errors were listed in my work plan: a Gantt chart that had the functionality of a Kanban chart embedded within it. Kanbans are developed at Toyota™ by the late Taiichi Ohno and they're part of what is called "Lean Product Development" of which "visual management" and "data visualisation" are key techniques (Sutherland 2015). These have become domestic household tools and are available even as mobile apps like the "Lean kit" for agile teams (www.Leankit.com). In explaining product management software, I've digressed. My deviation matters because these systems can omit digressions and cut short creative serendipity and time even to reflect. These project management tools promote strong independent problem solving, listing individual errors as temporal—to be solved depending on the time left on the project. Philosophy, unsurprisingly, did not appear on the Gantt/Kanban.

Isolating the Multi-part Polygon Error

The Panthera species exists in two continents. It's a grouped, multi-part polygon and for over one-third of the project's length, we used spatial data for Panthera shape files to test the design visuals and frontend development. The error pulled the two species together and was obscured by the issue of changing the testing perimeters before demoing the application. The team became familiar with the two shape files situated in South Africa and South America. These shape files became corrupted at the backend and therefore stopped being the primary user-journey. The shape became warped, stretching the shape range off the land and into the sea.

The marker should be centred within the shape, but it was marking Panthera in the middle of the Pacific, finding a centre point between two shapes rather than within the centre of one shape. The team had uncovered that markers were appearing in random places like at the edge of a polygon attaching to the first plotted point (see Fig. 6.1). Our frontend developer created an algorithmic rule to find the central point of a polygon. A square was drawn around the polygon to find its central point, and if that didn't appear in the centre, then the algorithm would retry again and again until it did so. Richard, the developer, said "*it's not a graceful way of doing it*". At first, the calculation was done at the frontend connected to our interface, but when that wasn't possible, the developers attempted it at the backend.

Two weeks before the end of the project, everyone thought the problem was solved. However, we had shifted to testing using a more detailed user-journey, that of an antelope chosen by the conservation organisation to demo to peers at a global summit. The centring problem worked fine on an antelope-species for a smaller and more localised mammal (its extinction is high).[5] The problem reappeared two months after the interface had been completed when it went for more testing with a group of mammologists. The Panthera shape file was no longer corrupted and the marker issue was easy to see, stretched across the equator, in the middle of the Pacific. With little time remaining on the project, the quick-fix was to remove markers altogether. All three partners agreed that there were other ways of identifying a polygon using the eye icon, colour and transparency. The system was error-friendly enough to cope with the deletion of markers and the designs allowed for markers to reappear in later versions of the tool. Accuracy in this GIS is the ability to measure degrees of uncertainty. GIS workers need to be able to interpret the levels of uncertainty and the "visual" noise of many multispecies, threats and interventions. As a reminder, Lucas points out, "*You can never pinpoint [and] say 'this is the boundary that this species can be found'*" (June 2012).

RETURNING TO THE PHILOSOPHICAL LOGICS OF THREAT-MAPPING

I've encountered three main challenges in this and other GIS design projects. They are the liveness of data, the changing perimeters for testing and the visibility and vistability of design issues. Continually changing perimeters also include new user-journeys and an unstable database with its fragmented taxonomies and shape files. The pace of the project exacerbated these challenges with little time to think fast about errors with slow theory because it's easy when doing digital work to get caught up in the speed of computation. What's out of kilter in the design testing phase is the reflexive literature aimed at designers. The literature aimed specifically at designers such as the quick tips and how-to guides of Scrum, Agile and Lean project management promote designing at speed, combining vocabulary from the sport, the battlefield and the marketplace (Sutherland 2015). However, more prevalent in the designing of this web application are the circular search logics, loops of problem-solving, telescopic viewing of flattened global data to a map and the aggregated point and line vectors. The project management texts prioritise temporal issues over spatial challenges.

When polygons group they remain as bubbles; they don't merge to make bigger bubbles—they stick together, or rather they group. In Sloterdijk's foam worlds, "individual bubbles are not drawn together into a single interrogative hyper-orb, but rather drawn together to make irregular hills" (I, 71). The problem is that when different kinds of bubbles are added to the foam (when they heap and layer), it's hard to distinguish groupings in an abstracted geometric flatland. In the interface, the foam and the bubbles are altogether—a problem of "spatial spreads" (foams) and "strong interdependencies" (bubbles) (ibid.). The bubble-foam issues of polygon-markers were obscured by changing data, warped shape files and changing testing perimeters (i.e. testing with the different Panthera and antelope).

Testing for errors needs to take place at all levels of zoom and should not rely too heavily on one particular viewpoint. Staying too long in one GIS or just in GISs altogether is to set up residency in a virtual "spatialised immune system" which is not preferential (TTM, 5). The vista-view of a GIS is a perspective often described in GIS literature as the "God-viewpoint" (Leszczynski 2009, 590). I was cautious of depending on this vistability when testing the app because GIS literature criticises this viewpoint. See geographer, Agnieszka Leszczynski, for a short note on research that stresses the "gods-eye-view" (2009, 590).[6] While this literature makes a valid point, it shouldn't inhibit the testing phase at all scales, otherwise it may become a gap in a visual-testing strategy.

WORKSHOP ASSESSING AND CITIZEN SCIENCE

Threat-mapping is a consensual workshopping process which relies on sharing GIS data brought back from the field. By digitising this process, results can be made visible to more people beyond the workshop group. Wider access to data extends the reach of possible participants and audiences. Therefore, the tool has citizen science potential. Searching this data does require skill to view and interpret such data. The skills of a data scientist and data visualiser are often intertwined in this work.

The consensual work done through discourse is similar to that discussed by Star and Griesemer (introduced in Chapter 3) when related to both the amateur and professional collectors of "live" specimens for museums (1989). My contribution to this form of writing is to explicate on the design process as a record of cooperative and consensual design-making from the perspective of a lead investigator of an interface design, interested in modes of reflexive, spherological knowing. I make no stake to ALL the design innovations related to this project or to the original idea, or novel use of original frontend or backend coding. Those belong to other team members and the group at large.

Adaptive Designing as a Way of Knowing Philosophy

By getting rid of markers we removed a way of making differences between shapes, but it's not an eradication of a positivist system. It just failed to make threats and interventions visible in a positivist way. Leszczynski argues that "representational knowledge-making" certifies the idea that ontologies "can be perceived and interrogated from any position, from any angle, and any scale" (2009, 590). However, Cartesian geometries like drawing tools are not *all* wrong when they help to support a non-Cartesian discourse, like evidencing the contamination of a river through media like reports, videos and photography.

Choosing to trace the Cartesian subject doesn't mean that positivist interpretations must follow. Reflecting on a Cartesian networked system through *Spheres* doesn't constitute a spatial turn away from Euclidean geometry, but rather it is a consistent recognition of spatial limitations. For instance, when right-angled polygons appear on the map it's more likely that this is because of human intervention (hence no rectangular drawing tools) like a national park boundary or a zoo enclosure (i.e. fences and walls), rather than a data-entry error. A shape with too many straight edges (a square) would raise concerns as to the data provenance. A rectangular drawing tool is more imprecise than drawing a sphere with an estimated radius in this GIS. The more right-angles in an area, the more human interventions there are, potentially. Rectangles are, however, very much present elsewhere in the GIS—in the application's window, menu boxes and the screen—the broader interfaces of GIS software. It's therefore not just a spherical and recursive tool. There are many network logics in the interface.

In threat-mapping, the rectangle is the crudest shape for drawing a threat or species. To push Sloterdijk's metaphor further, imagine foam that sits on top of a fast-running muddy river which appears dense in the flow of churned-up silt, but the foam on top is transparent when looked at up close. The spectral rainbows seem to bend over the co-isolated domes of the bubbling foam. In the web application, the colour and density of polygons are bright and vivid. However, they can still be difficult to distinguish in a thick stack of shape files. Millions of polygons on a map delineated by shaded areas and defined edges are a graphic designer's psychedelia. A density of data emphasises acute threats to multispecies. When tackling this issue, we worked in "scrum" mode (which means working on problems in individual "Sprints" of time, meeting regularly for brief catch-ups), adding more people to the team so that the psychedelia didn't lead to designer despair. As Sloterdijk explains:

> It is plain to see: at the edge of the phenomenal world, any thinking that stays phenomenological for too long turns into an internal water-color, which in the best of cases fades into non-technical contemplation.
>
> —INS, 94

Sloterdijk's point seems to make a dig at technical-reflection. Yet, when design remains overly technical, the phenomenological world of the designer may remain as the unopened black box of R&D issues. The design problems become insignificant to the computational and broader topical concerns of the application (threats to species). Surprisingly, design traditions for cartography like valuing the marker system wasn't a high priority for the environmental scientists. Often the design adapted to what was computationally doable rather than visually desirable.

While a user may ask the system to show "all" species and it may "hang" for a bit (the system shows a spinning timer, or in earlier versions, no error message at all) and finally "crash" (the program ends unexpectedly), the same may happen to design and development teams. Systems don't just crash and burn, designers and developers do too. Thankfully, the teams didn't crash because of sharing problems, be that in project management terms, or by eating together at the end of an enduring "sprint" as a voluntary activity of sharedness. This communality differs from the compulsory bread-making enforced on client-facing staff working on minimal wage or the "free" lunches brought to the employee's desk. These design practices try to encourage a particular type of collaboration and encourage non-stop 24/7 working in agoras that stress presenteeism. Jason and I often share stories of having worked in these design-specific environments.

And, what of the working environments of the computational environmental scientists? They share GIS software licenses, meaning they work around the clock, especially when a problem can take overnight to return a search query. The junior or interning scientists would wait their turn to do research, which often meant working into the night and early morning. The search enquirer waits awhile before this becomes apparent because the GIS will still try to perform a time-consuming task.[7]

A threat map is not a map of coherence; it's problematic, not a solution to shared working practices be that data analysis or consensual workshop mapping. GIS can be spaces for environmentalists to reside in so as to consensually workshop threats to species, but they are also for environmental scientists to momentarily reside in while waiting to go-in-to-the-field which requires funding and access rights. The answer to the problem of finding threats to species not only remains in the digital and the virtual sphere. To suggest this would imply that the virtual is more stable, static and robust than the threatened spaces that the environmental scientist hopes to highlight via GIS, which isn't the case.

The Importance of Embodied STS

This particular design problem meant orientating myself to the spatial work of a diverse group of environmental researchers, computational scientists, designers and developers. I didn't just need to get my head around the problem, but my body too. Getting my body around the problem meant not just bearing witness to a project (that initially sat uncomfortably with my queer-feminist politics)—that enacts a normative frame of reproduction and threat (a frame I "chose" to work within)—but it was also a way to recognise the norms I consistently design within.

For instance, this web application (like many digital tools) involve physically constrained work predominantly involving hand-to-eye coordination that relates to some mobile and remote working-in-the-field. The tool connects distributed people and spaces across the globe but is predominantly a desk-based tool for communal decision-making. There are feminists and queer researchers that would not choose to work on such a project, but I stay close here to the research of taxonomic practices by Susan Leigh Star (1989) and relational practices by Carla Hustak and Natasha Myers (2013).

The environmental work of this chapter involved decision-trees and taxonomic hierarchies which assemble species that can be identified by common and Latin names and categorised by sex and genus. Species-mapping is a feminist and a taxonomic issue that is well-rehearsed in environmental and STS literature by Donna Haraway (2016), Hugh Raffles (2010), Carla Hustak and Natasha Myers (2013), Susan Leigh Star (1989) and Stefan Helmreich (2009), but less is written about the translation work designers do in digital R&D.

At the time of doing this research, I was reading the first volume of *Spheres* which compounded the inescapable normative frames of threat- and species-mapping. Feminist and interactionist analyses of taxonomic science helped me to stay with the trouble of designing (Haraway 2016). Literature like those already listed can help future designers to remain self-aware in complex projects that rely on kinship relations to remember that interface design work contains environmental, proprietary and reproductive values that may differ from the designer's frame of reference. The literature, whether through reflexive commentary about eating onions (Star 1991; Star and Griesmer 1989), through to designing medical interfaces (Taylor et al. 2014) or cross-cutting analyses about reproduction and marriage (Strathern 1996), will alternatively provide a designer with literature that supports self-recognition as a necessary management of well-being.

I'll return briefly to Sloterdijk's nine-dimensional model which prioritises hand-to-eye issues over other bodily concerns (TTM, 4; and III, 462). Adjustments to sleep and global working patterns wouldn't feature in this model so it's not a philosophy for exploring embodied-user issues as I'll demonstrate. Sloterdijk outlines all dimensions except for one (which is absent) because the genesis of a human being cannot be constructed. They are chirotype (hand), phonotype (sound), erototype (cave dwelling), thermotype (warmth), erototype (jealousy and desire), ergotype (war and effort), thanatotope (of coexistence) and nomotype (law) (TTM, 4). The small issue relating to finding shapes on a map (i.e. polygon markers) was secondary to "searching" recursively. The "eye" was secondary to the "hand" because recursive searching reduces the repetitive use of a mouse or trackpad. There's no need to start a search over each time it isn't quite refined enough.

This nine-dimensional model belongs to one of three imaginary island scenarios. They include: *absolute islands*—space stations, *relative islands*—greenhouses and *anthropogenic islands*—spaces where humans can emerge (TTM, 4; and at length in III, 457–62). A GIS is an anthropogenic endeavour which can be considered an "absolute island" if understanding the data stored on a server connected to a graphical interface that packages up data as shapes on a map (TTM, 4). A global GIS will be a "relative island" if thought of as creating spaces for preserving a notion of nature as separate from human dwelling, like conservation parks and zoos as places of species residency.

Hustak and Myers entangle Darwin's choice of the orchid and insect bodies to rethink his evidence for natural selection. In this chapter, it's searching for threats and species that became the way of rethinking how environmental scientists evidence species-being-extinct, but GIS work also entangles technology markets and product design, and in doing so, has elevated designs' status in world problems, per say. How so-called "atmosphere designers" are tasked with some of the most depressive global issues affecting life-sustaining environments of dwelling directly linked to climate and extinction issues requires attention. To do so will help to understand the impact of design on multiple agencies and furthermore help contribute to interface design methods that prioritise care for the body and self-recognition.

When working on projects on a global level, normative frames abound. Designing at the global scale means staying close to the imbalances of normative life and to the protection of species. The way bodies are continually communicating the environmental damages of living temporally (rather than just spatially) in the anthropogenic era needs self-recognition. Here are three embodied examples. I've awoken in the middle of the night due to air pollution in London and I've had to walk because of shortness of breath when it's too polluted to run in the city. When I've heard an environmental scientist's voice via a Lync™ call (on another GIS project) I can hear them speak quietly (and somewhat weakly) closed-off in a room where they are, somewhere in Shanghai, sealed-off and air-tight away from temporarily dangerous levels of air pollution.

Some of my observations on this project weren't always groundbreaking or patentable—the treasures R&D promises to find. Instead, the broader embodied, environmental concerns help to set in context the lack of stress placed on mundane design issues like cartographic markers. In some respects, the portfolio of projects including this one constitutes the kind of design researcher that I am, which is often hard to define. I'm part of a foamed club of atmosphere designers trying to find designerly insights within the taxonomic frames of environmental identification.

When faced with difficult decision-making across various membership groups, what helped me was to remember the cooperation that exists between an extraordinarily diverse group of environmentalists who are embarking on these ever-complex atmospheric and climatic challenges—who are broadly also atmosphere designers.

Drawing on my interpretation of Star, atmosphere design as a feminist spherology is to do with working at the intersections of various membership groups including designers that read *Spheres*. The influential paper by Star, entitled "Power, Technology and the Phenomenology Conventions: On Being Allergic to Onions" ([1991] 2016) helps me to expand on understanding communities of practice (see Chapter 3) through the phrase "splitting selves" and "torque" (ibid., 263). Star draws on literature about bodily pain, notably the work of Elaine Scarry, an American academic of English literature (1985). Star combines Scarry with a range of literature on being a "stranger" drawing from sociologists Georg Simmel and Alfred Schutz (ibid., 283). From these ideas, Star develops the concept of "torque" to describe the problem of belonging in various membership groups. When a designer straddles communities of practice, torque is likely to be experienced. Torque isn't just experienced in ethnographic fieldwork, but in design that isn't necessarily ethnographic. Perhaps it is my fine art education that makes me feel like a stranger to design even though I've worked in the design industry and taught design for over two decades. I feel a stranger for all the reasons I've outlined in this book so far (communities of practice, torque, interdisciplinarity, health, gender, sexuality...).

Reading *Spheres* when designing creates another torque. Take, for instance, the way the two authors discuss Scarry's book on pain. For Sloterdijk the rhetorical power of torture and war shapes the narration of atmosphere design (III, 85–129). Star draws on Scarry's observation that Marie Curie's study of radiation did not extend to the impact it was having on her body as a point of self-recognition ([1991] 2016, 280–3). Combined with Star's insights about working at the intersections of various membership groups as another form of stress, these points on embodiment advance my argument for more research into the bodies of atmosphere designers—that goes beyond understanding body parts (limbs and organs) under stress, like repetitive stress injuries and eye strain, and towards ways in which global working is an "endochrine disrupter" (see Roberts 2007).

Through the process of taking Sloterdijk's atmosphere design into a contemporary computational ecology and design, I could have nearly drowned—to add a bit of narrative drama—if it wasn't for the ecological literature. While in the process of environmental designing, Helmreich (2009), Star (1989 [1995] 2016) and Raffles (2010) were a relief to read compared with *Spheres*. While Helmreich attends to the creatures of the ocean, Raffles attends to insects. While Sloterdijk argues for philosophy (CCR, xxxviii), I'd argue for something more situated, relational and located for designers.

While it's important to map threats and interventions, there are three necessary explications relating to the terms *species* (Haraway

[1999] 2008), *location* and *biodiversity* (Helmreich 2009)—made by critical anthropologists studying biology and microbiology research—that I found helpful in writing this chapter. Species, location and biological diversity are ways to differentiate things in space. Haraway and Helmreich figure species at the microbial level (like plankton) which aren't identifiable with the human eye. Helmreich, while building on Haraway, notes that not all microbial species are visible and parasitic. Therefore, their locations of origin "native or alien" are contested (ibid., 26). That is the dimensionality of species within other species.

Helmreich takes Marxist ideologies under the sea, whereas I've been recounting a different oceanic story about sunken cartographic markers. There's no single Marxist position in my tale because I'd hedge that there was enough variety of Capitalist, Liberalist, Marxist and Randist views of the economy representative in the design team and environmental scientists interviewed. Though I cannot be certain, however, because questions like these were out of project scope, but it's an intriguing question for future spherologists to ask.

Helmreich warns about "biodiversity" as a term proliferated by conservationists working at the quantitative scale, noting how nature becomes utilised in the following way: "…nature as a store of variety that might be measured and valued, on both ecological and economic terms" (2009, 110). Symbolising species, location and threats on a map is a representational pursuit of visual containment which has both economic and ecological value in GIS. It would be grandiose to suggest that the observations I've made could have been elicited only through design, or a philosophy like *Spheres*. Instead, my observations are reliant on STS and ethnographic reflections. This literature helps to relationally understand how Matthew *sees, feels* and *knows* about snowdrops in the Anthropocene while remaining closely attuned to species survival through modelling abstract, digital environmental data.

SOME CONCLUDING REMARKS ON THE DESIGNING OF FOAM WORLDS

The most critical misconception of this GIS design is to conceive threat-mapping as a complete project of which its wholeness can be seen if only we just had enough rendering capacity, enough vector-based shape files visible or a big enough vista (screen) to embrace the whole—to render more and crash less.

This interface design problem followed standard brand guidelines, conventional interview methods, well-established project management approaches and other design methods of development and testing. These methods create a sense of a project's wholeness but through logics of temporality rather than spatiality. However important it might be for a team's confidence and progress, the management literature can sometimes lead to a false sense of project certainty. Rather, the team's confidence was gained mainly through the insightful cooperation of all involved.

As I've noted in a discussion of *Spheres* and networks, there aren't so many linear rules to reflect on for a threat-mapping enquiry, but rather there are foam issues of recursive search. Data is shared among co-isolated groups of experts and volunteers, scientists and policymakers but also between those that segue those spaces, namely the threat-mapping assessors like Edward.

The error-checking phase was a particular moment for the design and development team, enabling us to pause midway through development so as to focus on click-through and live demos. While Lucas presented recursive search for identifying antelope and threats to the species in Korea, the UK-based team slept the day through. After resting, Jason, Richard and I met in the evening to celebrate over enormous pizza spheres in our local Italian restaurant, somewhere in London, south of the river, in an environment where fleas and moths, rats and mice, cats and dogs, squirrels and foxes, and more recently, rooks and green parrots coexist in abundance.

My intimacies and empathic relations to this project were not innocent—to parody Hustak and Myers—they were immersed in a personal investigation of *Spheres* and in finding a feminist spherological approach (2013, 92). Rather than extend a reading of *Spheres* as a model of philosophical certainty (based on individual parts of the body), I've rethought containment as an issue of design vistability (as embodied experience).

There appears to be enough space in many of the tropes of *Spheres* to consider them in a revised and relational way: like erring away from Sloterdijk's systems thinking of spatialised immune systems. In this chapter, I warn against designing computational tools as "island-making" activities, whereby digital tools replace going into the field. Web apps used in this way become hyped-orbs separated from the practices of working-in-the-field where foamed-worlds are disappearing.

At the broader relational scale, ecological "containment" problems in computational media are simultaneously a matter of embodiment and disembodiment—of sifting through the density of multispecies that sit atop of a map. Error-checking was a way to explain some of the challenges of atmosphere designing. Linking global and local species, threats and interventions helps to emphasise the depressed sense-making of atmosphere design done in converted farms and the glass offices of technology companies. That is, to *really* notice with eyes, ears, smell, taste and touch what is no longer alive nor well.

NOTES

1. All names have been anonymised for the environmental organisation. Everyone else mentioned has given permissions to be unanonymised.
2. Jason and I are partners. We've worked on several projects together and have lived together for over a decade. Our relationship was not known to the interviewees, but we were interviewed by my employer to discuss the conflicts of interest that could occur in this project. Conflicts resulted because of bringing work home, over-working and being overly critical. Benefits included an underpinning trust and a shared aesthetical language articulated not so much by words but by sight and tactility. Somewhere in between a "benefit" and a "risk" was the ability to work at speed.
3. For **climate** technologies see: air-conditioning units and radiators (III, 487); under-floor heating systems (ibid., 317); dwellings and self-design (ibid., 487); glass greenhouses and geodesic domes (II, 146; III, 318–30). For **atmosphere** technologies see: chlorine gas (ibid., 85); poison gas (ibid., 111); Zyklon A (ibid., 106); in warfare and torture (ibid., 97). For **air** technologies; see scents and air fresheners (ibid., 790); and climate art and museum spaces (ibid., 314).
4. *Spheres* offers for the GIS designer worldly representations of globes (i.e. III, 366), but this is where the illustrative materials fall short.
5. If a species has a small habitat and an even smaller spatial area that shows a threat, then the shape may not be greater in size to the marker. A threat assessor would need to find the threat by its exact latitude and longitude for the species.
6. Or, as Haraway calls "the God trick" (1991).
7. An error message was created to forewarn a GIS novice that searching for ALL isn't preferential.

References

Haraway, Donna. "The Biopolitics of Postmodern Bodies: Determinations of Self in Immune System Discourse." In *Feminist Theory and the Body: A Reader*, edited by Janet Price and Margrit Shildrick, 203–14. New York: Routledge, [1999] 2008.

———. *Simians, Cyborgs, and Women: The Reinvention of Nature*. London: Free Association, 1991.

———. *Staying with the Trouble*. Durham: Duke University Press, 2016.

Helmreich, Stefan. *Alien Ocean: Anthropological Voyages in Microbial Seas*. Berkeley, CA: University of California Press, 2009.

Hustak, Carla, and Natasha Myers. "Involutionary Momentum: Affective Ecologies and the Sciences of Plant/Insect Encounters." *Differences* 23, no. 3 (2013): 74–118.

Latour, Bruno. "Spheres and Networks: Two Ways to Reinterpret Globalization." *Harvard Design Magazine* 30 (2009): 138–44.

Law, John. *After Method: Mess in Social Science Research*. London: Routledge, 2004.

Leszczynski, Agnieszka. "Poststructuralism and GIS: Is There a 'Disconnect'?" *Environment and Planning D: Society and Space* 27 (2009): 581–602.

Raffles, Hugh. *Insectopedia*. New York: Pantheon Books, 2010.

Roberts, Celia. *Messengers of Sex: Hormones, Biomedicine and Feminism*. Cambridge: Cambridge University Press, 2007.

Scarry, Elaine. *The Body in Pain: The Making and Unmaking of the World*. Oxford: Oxford University Press, 1985.

Star, Susan Leigh. "Power, Technology, and the Phenomenology of Conventions: On Being Allergic to Onions." In *Boundary Objects and Beyond: Working with Leigh Star*, edited by Geoffrey C. Bowker, Stefan Timmermans, Adele Clarke, and Ellen Balka. Cambridge: MIT Press, [1991] 2016.

———. "Revisiting Ecologies of Knowledge: Work and Politics in Science and Technology (First Appeared in Titled Ecologies of Knowledge: Work and Politics in Science and Technology)." Edited by Geoffrey C. Bowker, Stefan Timmermans, Adele Clarke, and Ellen Balka, 13–47. Cambridge: MIT Press, [1995] 2016.

Star, Susan Leigh, and James R. Griesmer. "Institutional Ecology, 'Translations' and Boundary Objects: Amateurs and Professionals in Berkeley's Museum of Vertebrate Zoology, 1907–39." *Social Studies of Science* 19, no. 3 (1989): 387–420.

Strathern, Marilyn. "Cutting the Network." *The Journal of the Royal Anthropological Institute* 2, no. 3 (1996): 517–35.

Sutherland, Jeff. *The Art of Doing Twice the Work in Half the Time*. London: Business Books, Random House, 2015.

Taylor, Alex, Jasmin Fisher, Byron Cook, Samin Ishtiaq, and Nir Piterman. "Modelling Biology—Working Through (In-)Stabilities and Friction." *Computational Culture: A Journal of Software Studies* 1, no. 3 (2014). Accessed August 8, 2017. http://computationalculture.net/article/modelling-biology.

CHAPTER 7

Conclusions on Atmosphere Design

Abstract Creating a companion to *Spheres* was always going to be a cynical task, but to what extent can designers unpack the depressive sphere? Positioning *Spheres* theory at atmosphere designers means more than just adopting a cynical practice; rather, it involves considering design's inwardness as a condition of interiority related to the depressive sphere. Designers can do more than just bear witness to the depressive sphere by understanding their shared practices of design, and that includes the sociality of reading communally. *Spheres* is still relevant for today's temporal media issues and atmosphere designers tackling these issues can do so through the revised spherology of feminisms provided. Equipped with a renewed ethics of atmosphere and an awareness of creating activist-academia in relation to indexicality, readers may use this companion to go one step further and do some alchemy to extend the design-ins provided, *with* or *without Spheres*.

Keywords Atmosphere · Design-in · Feminisms · Media · Spherology

The initial enquiry of this book was to test if *Spheres* is a designer's philosophy. A companion about atmosphere design in the depressive sphere ensued, but that was through finding a spherology of feminisms. The format of a novel affords Sloterdijk an authorial individuality which is not without a loss to the reader. Artists and philosophers of celebrity status like Andy Warhol and Peter Sloterdijk get to record the world as they understand it, and in doing so, take a pop at it. *Spheres* does not aim

© The Author(s) 2019   143
A. Windle, *A Companion of Feminisms for Digital Design and Spherology*,
https://doi.org/10.1007/978-3-030-02287-7_7

to fragment its author; it intends to fragment the individual reader and control and tether newer disciplines (including design) to its philosophy of atmosphere. This assumes that disciplines in the smaller scale add to it, parasitically.

While reading Sloterdijk's cryptic autobiography, I hope the reader self-recognises the man in "Dasein". They may feel pain and then heal, but not in the couplings that Sloterdijk suggests—to find oneself through the annihilation of another. Rather than omit the fact that Sloterdijk was a Rajneeshee, this should help a designer to recognise the occult in design. A graphic designer knows how leading (like kerning) in typography does more than *just* manipulate space in letterform: it creates atmospheres. Consider the *obliqueness* of Helvetica, the clarity of Nimbus Sans, the #fontgate of Calibri and the co-isolated collectives of critical-speculative design.

The absences in Sloterdijk's trilogy are not up to the designer as art worker, translator or indexer to fix, but rather it is to recognise how they may be part of the system of indexicality that assumes this proposition. The environmental message to take away from my research is: "atmosphere designers" aren't fully equipped to help save the earth. Whenever positioned to do so the designer should first recognise that the author in *Spheres* is God-like and acting in a totalising way. Retorting to *Spheres* can lead to more discussion about theology. The theo-fable of this trilogy is that transpersonal meditation on the death of humanity is a luxury of writers and practitioners like Sloterdijk and not so readily afforded to academics today. So, I wayfared to Mexico to emphasise the clever stanzas, the quintillas of Sor Juana Inés de la Cruz, rather than her religious instruction and left it at that.

What I've read between and during R&D projects that are not budgeted for enables me to find new project directions and make different choices. R&D projects need to factor in time for reading so that budgetary demands don't squeeze the personal life out of a designer that does too much. It's not what academics in design or environmental science do for fun. It's caring for the globe and important labour. I hope that automated machines become an intelligent species because they may be the only things emotionally ready and resilient enough to learn about misogyny and survive long enough to see its extinction.

Rather than forget *Spheres*, I chose to write myself in (along with a few others along the way) like the practicing mystics, performance artists and designers, reparatively. I feel like I got to know each of them in some intimate way. The reader can figure the nude lido workers and the mystical saints that enliven up Sloterdijk's prose as something other than just lurid imagery. Focusing on the breaks, interruptions, errors and gaps in *Spheres* (as a strategy of reading) helps readers to rethink the confidence in Sloterdijk's writing and to show explicitly what's at stake. By breaking down grand narratives, or even grand digital designs, the ideas of *Spheres* are not shrugged off but taken seriously and in a feminist-queer-crip register. The method of writing consistently from the margins (to de-centralise spherology) is to take a different measure of the circumferences of foamed worlds. It's a method of unlearning the gendered and colonial traits of the sexual liberation found in *Spheres*.

If a critical reader cannot get Sloterdijk to consider a feminist reading as Huyssen urged, or cannot fully agree with Battersby's critique, or sit comfortably with the gender-neutrality of Babich and Wakeford or beat the shit out of a drumkit like Peters, then a different reading needs to take place. As the SCUM Manifesto of Solanas forewarns: Solana's writing came before the bullet that shot Warhol. To rhyme like Juana Inés de la Cruz, a small interstellar light warned out before that bullet that small meteorite shot out. My advice to readers when reading the countless other texts written in the same bombastic way as *Spheres* is to do a locative read, but to make sure it's not traumatic, obsessional, adhesive or destructive. It's important because the gaps in reading philosophy as a designer can be filled in anxiously.

Female ghosts call out from beneath the page, and that's why it's necessary to write-out all the missing women. It's healing and a reparative action, but it's also a Kynical game where Sloterdijk aims to win. There's a risk that the endurance and stamina it requires can only come from bodies that have been fragile. The other risk is to read it as a novel and to refrain from the academic research and not do the work it requires to understand and challenge its female explications. In the more frequent, darker and more depressive moments of reading *Spheres*, I've considered that if no one ever cited Sloterdijk again, it would not be soon enough. However, that then logically infers that to call out misogyny is to extend it. This thinking is futile. While some academic works should end things, most feminist writers go on after writing work like this to do other more positive things than dwell further in the depressive sphere, a trait I share with Sloterdijk.

Having said that, there are new challenges on the horizon. In 2017, NASA's Van Allen probes discovered that Very Low Frequency (VLF) radio, a submarine technology is creating an atmospheric bubble stretching out beyond the earth (Johnson-Groh 2017). Rather than consider this communication as a network, it would seem appropriate to consider this technical discovery as an atmosphere design. New bubble media could be researched through the spherology of feminisms provided in this companion.

Has Sloterdijk come full circle in his recent media ideas? When interviewed by Florian Inhauser for a think tank piece about social media by the Gottlieb Duttweiler Institute, Sloterdijk states that there will be a return to "quality" media in the next decade by lifting out of the "heated and hateful debate" which is deposed to the "lower end of the expression scale" (2018). In this interview, Sloterdijk seems to have forgotten that he's advocated for Kynicism, cathartic violence and faecal language for more than thirty years. Sloterdijk gets "his shit together" in 2018 by criticising the very same hateful speech acts he has used for three decades.

Academics need to guide without voyeurism so that readers tackling the depressive sphere (suggested in Chapter 5), can invent their own iterations of spherology without repeating the errors in *Spheres*. There need to be closer readings of literature—not just more enframing (in Chapter 4)—so that emerging scholars understand the importance of an ethics of atmosphere (in Chapter 3), know-how in relation to design credits (Chapter 5) and the finding of diverse indexicalities in what already exists (Chapter 4 and briefly 6). There need to be better places for practitioners to take care of their thoughts that impact their body (Chapter 2) so that the profession of design doesn't exile those that find community in feminisms. What echoes in the studio as agora and in globally mobile design work, is the eradication of intimate spaces, be they safe spaces, queer spaces or other spaces to be alone, to think and read uninterrupted (Chapters 1 and 2). This exile of spaces for introspection and time for autobiography in design is concerning.

I didn't stick to the theory, styles or structures of the philosophy of *Spheres* because designers need something else—like practical interventions. Design-ins were there in each chapter to help the reader when feeling exiled from *Spheres* such as data visualisations, design roughs and wireframes, etc. I started reading *Spheres* in various situations while designing, but I finished editing it in a hospital ward for women, awaiting surgery. The editing work became deeply connected to the act of finishing with *Spheres* and attending to my self-care. In one respect, I'm (in a sense) done with circling *Spheres* but not with a spherology of feminisms. This literature and these methods and modes of knowing will help a reader to pay close attention to themselves so that their own *being-in* isn't a *suspended into someone else's nothingness* (being-in *Spheres*).

Hopeful with parody and through a spherology of feminisms, atmosphere designers reading this companion can transit through the pages of *Spheres* so as to find space for reading and thinking autobiographically and to enquire how design is implicit in creating the depressive spheres. My companion revisits the ways a spherology of feminisms can be figured; but, this is not to leave *Spheres* behind. I'm not quite done with the depressive spheres and I'm not sure I can conclude optimistically because that's not the point, or where design is at right now.

Writers in design STS (rather than technoscience) generally leave out autobiography, but it doesn't matter how much you leave out: if you don't add it at some point, someone else might. My tale of intricate detail ends by stating loudly that investigating atmosphere design issues doesn't require having to get (~~shit~~) yourself together all of the time—though it does help to find the time to get stuff (feminisms) together, to persist in communicating *through*, *with* and *against* design in the depressive sphere of SCUM (annihilation).

References

Johnson-Groh, Mara. May 17, 2017. "NASA's Van Allen Probes Spot Man-Made Barrier Shrouding Earth." *NASA*. Accessed January 22, 2018. https://www.nasa.gov/feature/goddard/2017/nasas-van-allen-probes-spot-man-made-barrier-shrouding-earth.

Sloterdijk, Peter, and Florian Inhauser. January 22, 2018. *Social Media Is Followed by a Return to Quality*. Zurich: Gottlieb Duttweiler Institute. Accessed February 12, 2018. www.gdi.ch/de/Think-Tank/Trend-News/Auf-Soziale-Medien-folgt-die-Rueckkehr-zu-Qualitaet?sourceid=newsletter_20180208&utm_source=newsletter_20180208&utm_medium=E-Mail&utm_campaign=tribes18.

BIBLIOGRAPHY

Adorno, W.G. *Prisms*. Cambridge: MIT Press, 1955.

Ahmed, Sara. *Queer Phenomenology: Orientations, Objects, Others*. Durham: Duke University Press, 2006.

Ahmed, Sara. *Living a Feminist Life*. Durham: Duke University Press, 2017.

Akrich, Madeleine, Michel Callon, and Bruno Latour. "The Key to Success in Innovation Part I and II: The Art of Interessement." *International Journal of Innovation Management* 6, no. 2 (2002): 187–225.

Anthony, Kathryn H. *Designing for Diversity: Gender, Race, and Ethnicity in the Architectural Profession*. Minneapolis: University of Illinois Press, 2001.

d'Ávila, Santa Teresa. *The Life of St. Teresa of Avila by Herself*. Translated by J.M. Cohen. London: Penguin Books, [c.1536] 1957.

Babich, Babette. "Sloterdijk's Cynicism: Diogenes in the Marketplace." In *Sloterdijk Now*, edited by Stuart Elden, 17–36. Cambridge: Polity Press, 2012.

Back, Les. "Tape Recorder." In *Inventive Methods: The Happening of the Social*, edited by Celia Lury and Nina Wakeford, 245–61. Abingdon: Routledge, 2012.

Barad, Karen. "Posthumanist Performativity: Toward an Understanding of How Matter Comes to Matter." *Signs: Journal of Women in Culture and Society* 28, no. 3 (2003): 801–31.

Bates, Laura. 2012–. "Everyday Sexism Twitter Feed." Accessed January 21, 2016. https://twitter.com/EverydaySexism?ref_src=twsrc%5Egoogle%7Ctwcamp%5E-serp%7Ctwgr%5Eauthor.

Battersby, Christine. *The Phenomenal Woman: Feminist Metaphysics and the Patterns of Identity*. Cambridge: Polity Press, 1998.

———. "Her Body/Her Boundaries." In *Feminist Theory and the Body: A Reader*, edited by Janet Price and Margrit Shildrick, 341–59. Oxon: Routledge, 2008.

© The Editor(s) (if applicable) and The Author(s) 2019   149
A. Windle, *A Companion of Feminisms for Digital Design and Spherology*,
https://doi.org/10.1007/978-3-030-02287-7

Beattie, Tina. 2010. "Porete, a Forgotten Female Voice." *The Guardian*. Accessed January 26, 2016. www.theguardian.com/commentisfree/belief/2010/jun/07/marguerite-porete-mirror-simple-souls.

Bemberg, María Luisa (Director). *Yo, La Peor de Todas* (I, the Worst of All). Argentina: Crisalida Films, 1990.

Borson, Bob. 2011. "Life of an Architect." Accessed April 7, 2017. http://www.lifeofanarchitect.com/women-in-architecture/.

Breeze, Maddie. *Seriousness and Women's Roller Derby: Gender, Organisation, and Ambivalence*. Edinburgh: Palgrave Macmillan, 2015.

Bronkhurst, Judith. "1895: The Beggarstaffs' 'Annus Mirabilis'." *The Journal of the Decorative Arts Society 1890–1940*, no. 2 (1978): 3–13.

Brown, Juanita, and David Isaacs. *The World Cafe: Shaping Our Futures Through Conversations that Matter*. San Francisco: Barrett-Koehler Publishers, 2005.

Butler, Judith. *Excitable Speech: A Politics of the Performative*. New York: Routledge, 1997.

Butler, Octavia. *Lilith's Brood*. Also known as The Xenogenesis Trilogy. New York: Warner Books, [1987–1989] 2000.

Callon, Michel, and John Law. October 26, 2003. "On Qualculation, Agency and Otherness." Centre for Science Studies. Accessed January 29, 2018. http://www.lancaster.ac.uk/fass/resources/sociology-online-papers/papers/callon-law-qualculation-agency-otherness.pdf.

Cambrosio, Alberto, Peter Keating, and Andrei Mogoutov. "Mapping Collaborative Work and Innovation in Biomedicine: A Computer-Assisted Analysis of Antibody Reagent Workshops." *Social Studies of Science* 34, no. 3 (2004): 325–64.

Cixous, Hélène. "The Laugh of the Medusa." *Signs: Journal of Women in Culture and Society* 1, no. 4 (1976): 875–93.

Couture, Jean-Pierre. *Sloterdijk*. Cambridge: Polity Press, 2016.

de Beauvoir, Simone. *The Ethics of Ambiguity*. New York: Philosophical Library, 1948.

Deleuze, Gilles, and Felix Guattari. *A Thousand Plateaus; Capitalism and Schizophrenia*. London: Althone Press, 1988.

Denzin, Patrick. *Performance Ethnography: Critical Pedagogy and the Politics of Culture*. London: Sage, 2003.

Derrida, Jacques. *Margins of Philosophy*. Translated by Alan Bass, *Différence*. Chicago: Chicago University Press, 1972.

Descartes, Rene. *Renes Descartes: Principles of Philosophy*. Translated by V.R. Miller et al. Dordrecht: Springer (Reidel Publishing Company), 1644.

Dierkes-Thrun, Petra. "The Ecstatic Moment: Mysticism and Individualism in the Ecstasy of St. Teresa and Salomé." Accessed November 19, 2015. https://wildedecadents.wordpress.com/2012/12/17/the-ecstatic-moment-mysticism-and-individualism-in-the-ecstasy-of-st-teresa-and-salome/.

Dobrowolny, Wolfgang. *Ashram in Poona: Bhagwan's Experiment*. New York City: DRB, [1979] 1981.

Dieter, Michael. "The Virtues of Critical Technical Practices." *Differences: A Journal of Feminist Culture* 25, no. 1 (2014): 216–30.

Digital River. "Privacy Agreement for Dragon Nuance 5.0." Accessed August 8, 2017. https://shop.nuance.co.uk/DRHM/store?Action=DisplayDRPrivacy PolicyPage&SiteID=defaults&Locale=en_GB&ThemeID=22100&Env =BASE&eCommerceProvider=Digital%20River%20Ireland%20Ltd.

Dragon Nuance 5.0. "Dragon Installation and User Guide." Accessed July 28, 2017. http://www.nuance.co.uk/ucmprod/groups/corporate/@webenus/ documents/collateral/dns13_userguide.pdf#page240.

———. "Research Conducted in Conjunction with the Software." Accessed July 28, 2017. http://research.nuance.com.

———. "'Natural Language Speaking' Capabilities." Accessed July 28, 2017. http://www.nuance.co.uk/support/dragon-naturallyspeaking/index.htm.

Dunne, Anthony, and Fiona Raby. *Design Noir: The Secret Life of Electronic Objects.* Basel: August/Birkhäuser, 2001.

Dunne, Anthony. *Hertzian Tales: An Investigation into the Critical and Aesthetic Potential of the Electronic Product as a Post-Optimal Object.* London: Royal College of Art, 1997.

Elden, Stuart. *Sloterdijk Now.* Edited by Stuart Elden, 1–17. Cambridge: Polity, 2012.

Elsevier. 2017. *Gender in the Global Research Landscape.* Accessed June 24, 2017. https://www.elsevier.com/connect/ gender-and-science-resource-center#benchmark-report.

Foster Wallace, David. *Pale King.* London: Penguin, 2011.

Frampton, Edith. "Fluid Objects: Kleinian Psychoanalytic Theory and Breastfeeding Narratives." *Australian Feminist Studies* 19, no. 45 (2004). http://www.tandfonline.com/toc/cafs20/current.

Friedman, Sharon M. "Women in Engineering: Influential Factors for Career Choice." *Science Technology and Human Values* 2, no. 3 (1977): 14–16. http://journals.sagepub.com/home/sthOnline.

Gaiman, Neil. *American Gods.* New York: HarperCollins, 2017.

Gibbons, June. *Pepsi-Cola Addict.* New Horizon: Bognor Regis, 1982.

Haraway, Donna. *Simians, Cyborgs, and Women: The Reinvention of Nature.* London: Free Association, 1991.

———. "The Biopolitics of Postmodern Bodies: Determinations of Self in Immune System Discourse." In *Feminist Theory and the Body: A Reader,* edited by Janet Price and Margrit Shildrick, 203–14. New York: Routledge, [1999] 2008.

———. *Staying with the Trouble.* Durham: Duke University Press, 2016.

Häsing, Helga. *Du Gehst Fort, Und Ich Bleib Da: Gedichte Und Geschichten Von Abschied Und Trennung* (You Leave, and I Remain: Poems and Stories of Farewell and Separation). Germany: Fischer, 1992.

———. "Anthology." In *The Feminist Encyclopaedia of German Literature*, edited by Friederike Eigler and Susanne Kord, 20. Westport, CT: Greenwood, 1997.

———. *Main Never Partner Und Mean Kind* (My New Partner and My Child: Rejection, Acceptance, Solutions). Text-o-phon, 2001.

Häsing, Helga, and Ludwig Janus. *Ungewollte Kinder* (Unintended Pregnancy). Berlin: Rowohlt Taschenbuch, 1994.

Helmreich, Stefan. *Alien Ocean: Anthropological Voyages in Microbial Seas.* California: University of California Press, 2009.

Heidegger, Martin. *The Question Concerning Technology, and Other Essays.* Harper Perennial Modern Thought. New York and London: Garland, [1954] 1977.

Hexham, Irving, and Karla O. Poewe. *Understanding Cults and New Religions.* Michigan: Wm. B. Eerdmans, 1986.

———. *New Religions as Global Cultures.* New York: Perseus, 1997.

Heynen, Hilde. "Intervention in the Relations of Production, or Sublimation of Contradictions? On Commitment Then and Now." In *New Commitment: In Architecture, Art and Design* (Reflect 01). Rotterdam: NAi Publishers, 38–47, 2003.

Hine, Christine. *Virtual Ethnography.* London: Sage, 2000.

hooks, bell. *Feminist Theory: From Margin to Center.* Cambridge, MA: South End Press, [1984] 2000.

Houdart, Sophie. "Copying, Cutting and Pasting Social Spheres: Computer Designers' Participation in Architectural Projects." *Science Technology Studies* 21, no. 1 (2008): 47–63.

Hustak, Carla, and Natasha Myers. "Involutionary Momentum: Affective Ecologies and the Sciences of Plant/Insect Encounters." *Differences* 23, no. 3 (2013): 74–118.

Inés de la Cruz, Sor Juana. *Poems, Protest, and a Dream.* Translated by Margaret Sayers Peden. Introduction by Ilan Stevens. London: Penguin Classics, 1997.

Iso 999: 1996 Guidelines for the Content, Organization, and Presentation of Indexes. London: British Standards Institution (BSI).

Jaimes, Joel J. *Diversity at MIT.* 123, no. 3. Accessed August 8, 2017. http://tech.mit.edu/V123/N3/timeline.3f.html.

Jain, Sarah S. Lochlann. *Malignant: How Cancer Becomes Us.* Berkeley: University of California Press, 2013.

Johnson-Groh, Mara. May 17, 2017. "NASA's Van Allen Probes Spot Man-Made Barrier Shrouding Earth." *NASA.* Accessed January 22, 2018. https://www.nasa.gov/feature/goddard/2017/nasas-van-allen-probes-spot-man-made-barrier-shrouding-earth.

Jordán, Patricia Arriaga (Director). *Juana Inés.* Mexico: Canal Once, 2016.

Jungnickel, Kat. *Making Things to Make Sense of Things: DIY as Research and Practice.* London: Goldsmiths University Press, 2018.

Kang, Han. *The Vegetarian*. London: Portobello Books, 2007.

Karow, Yvonne. *Deutsches Opfer - Kultische Selbstausloschung Auf Den Reichsparteitagen Der Nsdap*. Berlin: Akademie Verlag, 1997.

Kember, Sarah. *Imedia: The Gendering of Objects, Environments and Smart Materials*. London: Palgrave Macmillan Pivot, 2015.

Kerouac, Jack. *On the Road*. London: Penguin, 1957.

Klein, Melanie. "Infantile Anxiety-Situations Reflected in a Work of Art and in the Creative Impulse." *International Journal of Psycho-Analysis* 10 (1929): 436–43.

Koskinen, Ilpo, John Zimmerman, Thomas Binder, Johan Redström, and Stephan Wensveen. *Design Research Through Practice: From the Lab, Field, and Showroom*. Burlington, MA: Morgan Kaufmann (Elsevier), 2011.

Lacan, Jacques. *The Seminar of Jacques Lacan 1972–1973*. Translated by Bruce Fink. Edited by Jacques-Alain Miller. Vol. XX, *On Feminine Sexuality. The Limits of Love and Knowledge* (Encore). London: W. W. Norton, 1998.

Latour, Bruno. *Science in Action*, Milton Keynes: Open University Press, 1987.

———. 2008. "A Cautious Prometheus? A Few Steps Toward a Philosophy of Design (With Special Attention to Peter Sloterdijk)." Accessed August 8, 2017. http://www.universal-publishers.com/book. php?method=ISBN&book=1599429063.

———. "Spheres and Networks: Two Ways to Reinterpret Globalization." *Harvard Design Magazine* 30 (2009): 138–44.

Latour, Bruno, and Peter Weibel, eds. "From Realpolitik to Dingpolitik Or How to Make Things Public", In *Making Things Public*, 40–61. Cambridge: MIT Press, 2005.

Latour, Bruno, Pablo Jensen, Tommaso Venturini, Sébastian Grauwin, and Dominique Boullier. "The Whole Is Always Smaller Than Its Parts—A Digital Test of Gabriel Tardes' Monads." *The British Journal of Sociology* 63, no. 4 (2012): 590–615.

Lave, Jean, and Etienne Wenger. *Situated Learning: Legitimate Peripheral Participation*. Cambridge: Cambridge University Press, 1992.

Law, John. *After Method: Mess in Social Science Research*. London: Routledge, 2004.

Law, John. "Modes of Knowing: Resources from the Baroque." In *Modes of Knowing*, edited by John Law and Evelyn Ruppert, 17–58. Mattering Press, 2016. https://www.matteringpress.org/.

Leszczynski, Agnieszka. "Poststructuralism and GIS: Is There a 'Disconnect'?" *Environment and Planning D: Society and Space* 27 (2009): 581–602.

Lindström, Kristina, and Åsa Ståhl. "Working Patches." *Studies in Material Thinking* 7, no. 4 (2012): 1–17. AUT University. https://www.materialthinking.org/sites/default/files/papers/SMT_V7_P4_Lindstr%C3%B6mSt%C3%A5hl_0.pdf

Linton, S. *Claiming Disability: Knowledge and Identity*. New York: New York University Press, 203, 1998.

Lupton, Ellen, J., and Abbott Miller. *Design, Writing, Research: Writing on Graphic Design*. London: Phaidon, 1996.

Lury, Celia, and Nina Wakeford. *Inventive Methods: The Happening of the Social*. London: Routledge, 2012.

MacLeod, Scott. 2016. "Global Trouble (Interview with Judith Butler)." *Cairo Review*. Accessed August 8, 2017. https://www.thecairoreview.com/q-a/global-trouble/.

Malpass, Matt. *Critical Design in Context: History, Theory, and Practices*. London: Bloomsbury, 2017.

Marres, Noortje. "May the True Victim of Defacement Stand Up! On Reading the Network Configurations of Scandal on the Web." In *Making Things Public*, edited by Bruno Latour and Peter Weibel, 486–89. Cambridge: MIT Press, 2005.

Marres, Noortje, and Richard Rogers. "Recipe for Tracing the Fate of Issues and Their Publics on the Web." In *Making Things Public*, edited by Bruno Latour and Peter Weibel, 922–33. Cambridge: MIT Press, 2005.

Meadows, Alice. "The Global Gender Gap: Research and Researchers." Accessed August 8, 2017. https://scholarlykitchen.sspnet.org/2017/03/08/the-global-gender-gap-research-and-researchers/.

Milne, Elisabeth-Jane. "Saying 'No' to Participatory Video: Unravelling the Complexities of (Non)Participation." Edited by C. De Lange, E. J. Milne, and N. Mitchell, 257–68. Lanham: AltaMira, 2012.

Morin, Marie-Eve. "The Coming-to-the-World of the Human Animal." In *Sloterdijk Now*, edited by Stuart Elden, 77–95. Cambridge: Polity Press, 2012.

Müller-Doohm, Stefan. *Adorno: A Biography*. Malden, MA: Polity Press, 2005.

Müeller von der Haegen, Gesa. 2011. "Container Architecture for Nrw-Forum Düsseldorf Exhibition Catalogue." *Arch Daily*. Accessed August 28, 2017. http://www.archdaily.com/131885/exhibition-container-architecture-at-the-nrw-forum-museum-in-duesseldorf.

Musil, Robert. *The Man Without Qualities*. Edited by Burton Pike, with Sophie Wilkins. London: Picador, 1966.

Myerson, Jeremy, and Philip Ross. *Space to Work: New Office Design*. London: Lawrence King Publishing, 2006.

Nussbaum, Martha C. *Cultivating Humanity: A Classical Defense of Reform in Liberal Education*. Cambridge: Cambridge University Press, 1997.

———. "Aristotelian Social Democracy." In *Liberalism and the Good*, edited by R. Bruce Douglass, Gerald M. Mara, and Henry S. Richardson, 289. London: Routledge, 1990.

Oliveira, Carlos, Peter Sloterdijk, and Carl Hanser. *Selbstversuch: Ein Gespräch Mit Carlos Oliveira*. München: Verlag, 1996.

Orr, Jackie. *Panic Diaries: Genealogy of Panic Disorder*. Durham: Duke University Press, 2006.

Orr, Susan, Margo Blythman, and Joan Mullin. "Textual and Visual Interfaces in Design Education." *Art, Design & Communication in Higher Education* 3, no. 2 (2004): 75–80.

Paglia, Camilla. "Feminism, In Conversation with Camilla Paglia." *The Battle of Ideas.* Accessed February 12, 2018. http://archive.battleofideas.org.uk/2016/session/11488#.WocWJZOFjiw.

Parry, Eugenia. *Joel-Peter Witkins. Catalogue 55.* London: Phaidon Press, 2001.

Perloff, Marjorie. 2002. "Dada Without Duchamp / Duchamp Without Dada: Avant-garde Tradition and the Individual Talent." *Electronic Poetry Center.* Accessed December 8, 2017. http://epc.buffalo.edu/authors/perloff/dada.html.

Pérez-Bustos, Tania. "El tejido como conocimiento, el conocimiento como tejido: reflexiones feministas en torno a la agencia de las materialidades" (O tecido como conhecimento, o conhecimento como tecido: reflexões feministas sobre a agência das materialidades) [Weaving as Knowledge, Knowledge as Weaving: Feminist Reflections on the Agency of Materialities]. *Revista Colombiana de Sociología,* 39, no. 2 (2016): 163–82. https://revistas.unal.edu.co/index.php/recs/article/view/58970.

Pink, Sarah, László Kürti, and Ana Isabel Afonso. *Working Images: Visual Representation in Ethnography.* New York: Routledge, 2004.

Pitt, Joshua. April 10, 2017. "Silenced Issues 3: Silenced Issues in Academic Publishing." *In Backchannels.* Accessed September 3, 2017. http://www.4sonline.org/blog/post/silenced_issues_3_silenced_issues_in_academic_publishing.

Poewe, Karla O. *Universal Male Dominance: An Ethological Illusion.* Amsterdam: Elsevier, 1980.

Poewe, Karla O. *My Apprenticeship: An Intellectual Journey.* Calgary: Vogelstein Press, 2018.

Porete, Margareta [Or Unknown French Mystic of the Thirteenth Century]. *The Mirror of Simple Souls.* Translated by Clare Kirchberger. London: Burns Oates and Washbourne Ltd., 1927.

Posavec, Stefanie. "Literary Organism." Accessed January 26, 2016. http://www.stefanieposavec.co.uk/personal/#/writing-without-words/.

Proust, Marcel. *Search of Lost Time. Swann's Way.* London: Penguin, 1913.

Raby, Fiona. "Interview." In *Darkitecture: Learning Architecture for the Twenty-First Century,* edited by Gerrard O'Carroll, 84–87. London: Two Little Boys, 2012.

Raffles, Hugh. *Insectopedia.* New York: Pantheon Books, 2010.

Raitt, Jill. "Laudem Caroli: Renaissance and Reformation Studies for Charles G. Nauert: Sixteenth Century Essays and Studies," Xlix no. 5. In *The Two Spiritual Directors of Women in the Sixteenth Century: St. Ignatius Loyola and St. Teresa of Avila,* edited by James V. Mehl. Missouri: Thomas Jefferson University Press, 1998.

Rajneesh, Bhagwan Shree. *Chapter Two: The Greatest Discovery There Is*. Osho Online Library. Accessed January 23, 2018. http://www.osho.com/iosho/library/read-book/online-library-sex-condemned-joke-8251a6bd-c62?p=84c4f2236f2b261789ef7a6d9dd842b7.

Raschke, Carl. "Peter Sloterdijk as First Philosopher of Globalization." *Journal for Cultural and Religious Theory* 12, no. 3 (2013): 1–19.

Rashof, Sascha. *Designing Place—Topologies of Maker Labs*. Doctoral thesis, Goldsmiths, University of London [Thesis], 2016.

Roberts, Celia. *Messengers of Sex: Hormones, Biomedicine and Feminism*. Cambridge: Cambridge University Press, 2007.

Rosner, Daniela, *Critical Fabulation: Reworking the Methods and Margins of Design*, Cambridge: MIT Press, 2018.

Rossmann, Andreas. June 30, 2011. "The Hamburger of Architecture." *Style Park*. Accessed August 30, 2015. www.stylepark.com/en/news/the-hamburger-of-architecture/322093.

Rossiter, Margaret W. "The Matthew Matilda Effect in Science." *Social Studies of Science* 23, no. 2 (1993): 325–41.

Roudinesco, Élisabeth. 2015. "The Sculptural Iconography of Feminine Jouissance: Lacan's Reading of Bernini's Saint Teresa." Accessed January 27, 2016. www.lacanschool.org/events/the-sculptural-iconography-of-feminine-jouissance-lacans-reading-of-berninis-saint-teresa/.

Roy, Deboleena. "Feminist Approaches to the Enquiry in the Natural Sciences: Practices in the Lab." In *Handbook of Feminist Research: Theory and Praxis*, edited by Sharlene Nagy Hesse-Biber, 312–29. London: Sage, 2011.

Rughani, Pratap. *The Dance of Documentary Ethics*. Edited by Brian Winston, 98–109. British Film Institute, London: Palgrave Macmillan, 2013.

Sacks, Oliver. October 19, 1986. "Bound Together in Fiction and Crime." *New York Times*. Accessed August 5, 2017. http://www.nytimes.com/1986/10/19/books/bound-together-in-fantasy-and-crime.html?pagewanted=all.

Saldanha, Charlotte, Dominic Smith, and Amanda Windle. 2015. "Audience Development Via Digital Means. Voice of Culture." *Structured Dialogue Between the Arts and Culture Sector and the EU Commission*. Accessed October 15, 2018. http://ualresearchonline.arts.ac.uk/8666/.

Scarry, Elaine. *The Body in Pain: The Making and Unmaking of the World*. Oxford: Oxford University Press, 1985.

Schön, Donald A. *The Reflective Practitioner: How Professionals Think in Action*. New York: Basic Books, 1983.

Sleeman, Cath, and Nina Cromeyer Dieke. November 13, 2015. *Antimicrobial Resistance Across Europe*. Accessed June 22, 2017. http://www.nesta.org.uk/blog/fight-against-antimicrobial-resistance-across-europe.

Sloterdijk, Peter. *Critique of Cynical Reason*. Translated by Michael Eldred. Edited by M. Eldred. Minneapolis: University of Minnesota Press, [1983] 1987.

———. "Atmospheric Politics." In *Making Things Public: Atmospheres of Democracy*, edited by Bruno Latour and Peter Weibel, 944–57. Cambridge: MIT Press, 2005.

———. "Spheres Theory: Talking to Myself About the Poetics of Space." *Harvard Design Magazine* 30 (2009): 1–8.

———. *The Aesthetic Imperative: Writings on Art*. Translated by Karen Margolis. Edited by Peter Weibel. Cambridge: Polity, [2014] 2017.

———. *Spheres, Bubbles: Microspherology*. Translated by Wieland Hoban. Vol. I. Los Angeles: Semiotext(e), 1998.

———. *Spheres, Globes: Macrospherology*. Translated by Wieland Hoban. Vol. II. Los Angeles: Semiotext(e), 1999.

———. *Spheres, Foams: Plural Spherology*. Translated by Wieland Hoban. Vol. III. Los Angeles: Semiotext(e), 2004.

———. *Spharen Iii: Shäume Plurale Sphärologie*. Frankfurt: Suhrkamp, 2004.

———. "Inspiration." *Ephemera: Theory and Politics in Organization* 9, no. 3 (2009): 242–51.

———. *Terror from the Air*. Translated by S. Corcoran. Los Angeles: Semiotext(e), 2009.

Sloterdijk, Peter, and Florian Inhauser. January 22, 2018. *Social Media Is Followed by a Return to Quality*. Zurich: Gottlieb Duttweiler Institute. Accessed February 12, 2018. www.gdi.ch/de/Think-Tank/Trend-News/Auf-Soziale-Medien-folgt-die-Rueckkehr-zu-Qualitaet?sourceid=newsletter_20180208&utm_source=newsletter_20180208&utm_medium=E-Mail&utm_campaign=tribes18.

Sloterdijk, Peter, and Gesa Mueller von der Haegen. "Instant Democracy: The Pneumatic Parliament(R)." In *Making Things Public: Atmospheres of Democracy*, edited by Bruno Latour and Peter Weibel, 952–58. Cambridge: MIT Press, 2005.

Solanas, Valerie. *SCUM Manifesto*. London: Verso, [1967] 2004.

Solnit, Rebecca. *When Men Explain Things to Me*. London: Haymarket Books, 2015.

Stackelberg, Roderick. "Reviewed Work: Deutsches Opfer: Kultische Selbstauslöschung Auf Den Reichsparteitagen Der Nsdap by Yvonne Karow." *Central European History* 32, no. 4 (1999): 493–95.

Star, Susan Leigh. "Power, Technology, and the Phenomenology of Conventions: On Being Allergic to Onions." In *Boundary Objects and Beyond: Working with Leigh Star*, edited by Geoffrey C. Bowker, Stefan Timmermans, Adele Clarke, and Ellen Balka. Cambridge: MIT Press, [1991] 2016.

———. "Misplaced Concretism and Concrete Situations: Feminism, Method and Information Technology." In *Boundary Objects and Beyond: Working with Leigh Star*, edited by Geoffrey C. Bowker, Stefan Timmermans, Adele Clarke, and Ellen Balka, 143–71. Cambridge: MIT Press, [1994] 2016.

———. "Revisiting Ecologies of Knowledge: Work and Politics in Science and Technology (First Appeared in Titled Ecologies of Knowledge: Work and Politics in Science and Technology)." Edited by Geoffrey C. Bowker, Stefan Timmermans, Adele Clarke, and Ellen Balka, 13–47. Cambridge: MIT Press, [1995] 2016.

Star, Susan Leigh, and James R. Griesmer. "Institutional Ecology, 'Translations' and Boundary Objects: Amateurs and Professionals in Berkeley's Museum of Vertebrate Zoology, 1907–39." *Social Studies of Science* 19, no. 3 (1989): 387–420.

Stein, Gertrude, *The Autobiography of Alice B. Toklas.* London: Penguin Modern Classics, [1933] 2001.

Stengers, Isabelle. *Power and Invention: Situating Science.* Translated by P. Bains. Minneapolis: Minnesota Press, 1997.

———. "Introductory Notes on an Ecology of Practices." *Cultural Studies Review* 11, no. 1 (2005a): 183–96.

———. "The Cosmopolitical Proposal." In *Making Things Public*, edited by Bruno Latour and Peter Weibel, 994–1004. Cambridge: MIT Press, 2005b.

Strathern, Marilyn. "Cutting the Network." *The Journal of the Royal Anthropological Institute* 2, no. 3 (1996): 517–35.

Suchman, Lucy. *Human-Machine Configurations: Plans and Situated Actions.* Cambridge: Cambridge University Press, [1987] 2007.

Sutherland, Jeff. *The Art of Doing Twice the Work in Half the Time.* London: Business Books, Random House, 2015.

Taylor, Alex, Jasmin Fisher, Byron Cook, Samin Ishtiaq, and Nir Piterman. "Modelling Biology—Working Through (In-)Stabilities and Friction." *Computational Culture: A Journal of Software Studies* 1, no. 3 (2014). Accessed August 8, 2017. http://computationalculture.net/article/modelling-biology.

Teil, Genevieve, and Bruno Latour. 1995. "The Hume Machine: Can Association Networks Do More Than Formal Rules? Constructions of the Mind." Accessed April 15, 2009. www.stanford.edu/group/SHR/4-2/text/teil-latour.html.

Treister, Suzanne. 2007–2008. "Alchemy Project." Accessed January 21, 2016. www.suzannetreister.net/ALCHEMY/ALCHEMY.html.

Van Tuinen, Sjoerd. "Air Conditioning Spaceship Earth: Peter Sloterdijk's Ethico-Aesthetic Paradigm." *Environment and Planning D: Society and Space* 27 (2009): 105–18.

———. "A Thymotic Left? Peter Sloterdijk and the Psychopolitics of Ressentiment." *Symploke* 18, no. 1–2 (2010): 47–64.

Verran, Helen, and Brit Ross Winthereik. "Innovation with Words and Visuals: A Baroque Sensibility." In *Modes of Knowing*, edited by John Law and Evelyn Ruppert, 197–223. Mattering Press, 2016. https://www.matteringpress.org/.

Wakeford, Nina. 2011. "Replacing the Network Society with Social Foam: A Revolution for Corporate Ethnography." Accessed August 8, 2017. http://epicpeople.org/wp-content/uploads/2014/09/Wakeford_repla.pdf.

———. "Don't Go All the Way: Revisiting 'Misplaced Concretism'." In *Boundary Objects and Beyond: Working with Leigh Star*, edited by Geoffrey C. Bowker, Stefan Timmermans, Adele Clarke, and Ellen Balka, 69–84. Cambridge: MIT Press, 2016.

Walker, Barbara. *Women's Encyclopedia of Myths and Secrets.* New York: HarperCollins, 1983.

Wallace, Marjorie. *The Silent Twins.* Edited by Jon (Director) Amiel. British Broadcasting Corporation, 1986.

———. "The Tragedy of a Double Life." *The Guardian.* Accessed August 8, 2017. https://www.theguardian.com/society/2003/jul/13/health.lifeandhealth.

Walsh, Roxy. "2011. RSVP Revisited. (Rsvp at Annika Sundvik Gallery, New York 1994 Temple Bar Gallery," Dublin. Accessed January 26, 2016. http://www.roxywalsh.com/page18.htm.

Warwick, G. *Bernini: Art as Theatre.* New Haven, CT, and London: Yale University Press, 2012.

Way, Maclain and Chapman (Directors). *Wild Wild Country*, Duplass Brothers Productions, Netflix Original Series. 2018.

Webb, Sharon. 2015. "Requirements and National Digital Infrastructures: Digital Preservation in the Humanities." Accessed August 5, 2017. http://sro.sussex.ac.uk/57120/.

Whitehead, C. *The Intuitionist.* London: Granta/Anchor Books, 1999.

Windle, A. "Automation and Design for Prevention: Fictional Accounts of Misanthropic Agency from the Elevator (Lift) to the Sexbot (Chatbot)." *Technoetic Arts* (2014): 91–106.

———. June 23, 2016. "A Personal Plea from Europe's Polar Bears" In *New Statesman Tech.* http://tech.newstatesman.com/enterprise-it/european-digital-research-funding.

Winterson, Jeanette, *Art Objects: Essays on Ecstasy and Effrontery.* London: Vintage, 1996.

Wittkower, R. *Bernini: The Sculptor of the Roman Baroque.* London and New York: Phaidon. ([1955] 1997) In *Art and Architecture in Italy 1600–1750: II High Baroque*, edited by J. Connors and J. Montague, rev. New Haven, CT, and London: Yale University Press, [1958] 1999.

INDEX

© The Editor(s) (if applicable) and The Author(s) 2019
A. Windle, *A Companion of Feminisms for Digital Design and Spherology*,
https://doi.org/10.1007/978-3-030-02287-7